CAMBRIDGE COUNTY GEOGRAPHIES

General Editor: F. H. H. GUILLEMARD, M.A., M.D.

# RADNORSHIRE

*Cambridge County Geographies*

# RADNORSHIRE

by

## LEWIS DAVIES
(of Cymmer)

With Maps, Diagrams and Illustrations

Cambridge:
at the University Press

1912

CAMBRIDGE UNIVERSITY PRESS
Cambridge, New York, Melbourne, Madrid, Cape Town,
Singapore, São Paulo, Delhi, Mexico City

Cambridge University Press
The Edinburgh Building, Cambridge CB2 8RU, UK

Published in the United States of America by Cambridge University Press, New York

www.cambridge.org
Information on this title: www.cambridge.org/9781107691414

First published 1912
First paperback edition 2013

*A catalogue record for this publication is available from the British Library*

ISBN 978-1-107-69141-4 Paperback

# PREFACE

I HAVE to thank Mrs Edmund J. Jones, "Fforest Legionis," Glynneath, for kind aid in the subject matter of Ecclesiastical Architecture; Mr Roger Howel, Bryncoch, for facts of interest in connection with the geology of the county; and Mr T. R. Thomas, Cumtwrch, for his valuable assistance in correcting the proofs.

L. D.

*July* 1912

# CONTENTS

# ILLUSTRATIONS

## MAPS

The illustrations on pp. 72, 97, 111, 112, and 149 are from photographs by Mr W. M. Dodson, Bettws-y-Coed; the remainder are from photographs by Mr P. B. Abery, Builth Wells; the maps on pp. 15, 21, and 120 are from sketches by the author; that on p. 49 is from a sketch kindly supplied by Dr J. E. Marr; that facing p. 116 is from a drawing by Mr C. J. Evans.

# 1. The Peninsula of Fferllys. The Place of Radnor in Wales. The Name—its Origin and Meaning.

A glance at the map of England and Wales shows that two of our largest rivers—the Severn and the Wye—not only rise within a short distance of each other on the eastern slope of Plynlimmon, but that after wandering sixty miles apart they meet again before entering the sea.

The land included by these streams is thus really a peninsula, and was known to the ancients by the collective name of Fferllys, or Fferyllwg. The greater part of Radnorshire lies within its north-west portion, occupying the same tract as the lordships of Melenydd and Elvael in far-off times. Another district—that of Gwrtheyrnion—a mountainous tract to the west of the Wye—was added to Melenydd and Elvael to form the complete county when its present boundaries were fixed.

This region of hill and dale has always been famous. Its mountainous character has served as an excellent defence in all ages. Its position, too, as a debatable borderland between Dyfed on the one side and Mercia on the other, made its possession, especially in medieval times, a matter of great import. All the roads and tracks of Mid-Wales converged on the fords of Wye, and whoever dominated them dominated the country on both sides.

Celt, Roman, and Norman successively mastered and owned the beautiful valleys of Upper Fferllys, and each

A Radnorshire Moorland Scene

has left very conspicuous marks of his occupation. The Saxon was not uniformly victorious—his boundary varied from age to age. Now he advanced to the Wye itself, and now he had to retreat to the Severn, until at last the great Offa fixed the boundary that is still connected with his name. The Dane also raided the district, but did not succeed in making it his permanent abode.

Although this mountain region has a long history, the county of Radnor, as we now know it, is a modern division, dating no further back than 1536, when at the same time, and under similar circumstances as Brecon, Monmouth, Montgomery, and Denbigh, it was created out of the unshired lands of the Borders, or Marches.

Vale of New Radnor

Pembroke and Glamorgan were counties before the fall of the last Llewelyn, and in 1284 Edward I by the Statute of Rhudd-lan created six others—Flint, Carnarvon, Anglesey, Merioneth, Cardigan, and Carmarthen; but the remainder of Wales was for two and a half centuries an unhappy territory, where legalised oppressors, the Lords-Marchers, to the number of about

140, held the destinies of life and death over the wretched people, ruling them by the strength of the sword, in virtue of the "Jura Regalia,"—a martial law of the sternest kind.

Henry VIII by his Act of Union in 1536 swept all this away by adding a number of the Lordships to the existing neighbouring English and Welsh shires, and forming the remainder into five new counties, of which, as we have seen, Radnor was one.

Although the name Radnor, as applied to the county, is comparatively modern, the name as applied to the chief mountains of the district, Radnor Forest, and to the villages of Old and New Radnor, is a very old one, *New* Radnor even being at least as old as the Domesday Book. The meaning of the word has long been a puzzle to scholars, but the interpretation most generally accepted derives it from the Anglo-Saxon *rade* = "a road" and *nore* = "narrow," making it to mean "the land of mountain tracks," which it most certainly was.

The origin of the Welsh name—Maesyfed—is also doubtful, for although *maes* = "a field" is definite enough, there is no satisfactory explanation of the other portion of the word, the various authorities offering as many explanations, ranging from *yfed* = "to drink" and *Hyfaidd* = "a Welsh chieftain," to *y fedw* = "the birch tree." The last mentioned has at least the merit of being plausible, especially if compared with the Welsh names of other border towns, e.g. Pengwern (Shrewsbury), "the head of the alders," Tre-ffawydd (Hereford), "the town of the beeches," and Celyn (Clun), "the town of holly."

## 2. General Characteristics.

Radnorshire has no coal, and is thus likely always to remain a pastoral county. She has, moreover, no rivers of commercial utility and no sea-coast. Indeed, it is the Welsh county most remote from the sea, and so mountainous is its general character that it has earned the

A Radnorshire Farm

name of "rugged Radnor," though certainly not from the poverty of its elevated slopes. For the crisp turf of the Radnor heights has made it renowned among the sheep-breeders of the Principality, while its sheep-walks are in great request by stock raisers at all times.

Heather is everywhere plentiful in the uplands, and as a consequence grouse abound. The county in ancient days contained no less than four royal forests, and its sparse population, and the great extent of its moorland, made it peculiarly suitable for the chase.

Its valleys, though generally narrow, are of extreme beauty, "no turf being greener or fresher," and "no streams being clearer, and more buoyant" than those which form the groundwork of the dales around the foot of the Great Forest. Byron and Shelley felt their charm, for the former was a constant visitor at Kensham, near Presteign, and Shelley took up his abode in the Elan Valley during a period of his fitful career. There is an old distich which says,

> "Blessed is the eye
> 'Twixt Severn and Wye,"

in allusion to the acknowledged beauty of this Mesopotamia of the Welsh Border, and Shakespeare himself has sung the same theme when in *King Lear* (i. i. 65) he describes Fferllys as

> "With shadowy forests and with champains rich'd,
> With plenteous rivers and wide-skirted meads."

Radnor is becoming increasingly the playground of South Wales, not only for those who can rent a shooting or fishing, but for all classes.

Anglers come in great numbers, for although most of the streams are preserved, there are still many left where fishing may be had free, or on payment of a small sum. Latterly, the well-stocked chain of lakes formed in the Elan district for the domestic needs of Birmingham have greatly added to the fishing value of the county, and anglers from far afield regularly visit them to fish.

Radnorshire is very rich in antiquities of all kinds. The county indeed may be summed up as a curious mixture of the very old and the very new ; where Offa's Dyke, with a story of 1200 years, lies in the path

The Pump House, Llandrindod

of the Birmingham Aqueduct of yesterday, and where Roman camps, castle-mounds, and prehistoric graves are lit up every evening by the electric light of Llandrindod Common.

Lying, as it did, right on the Mercian border, and, moreover, being halfway between North and South Wales, Radnor, as may be surmised, was constantly the battleground of the contending forces. But, throughout the ages of fierce strife, it remained Welsh in tongue and character until within a century ago, when in times of peace it forgot the Cymric language. It is the only Welsh county that has entirely done so, but it can never be other than Cymric in character as long as its place-names survive, and its traditions of Llewelyn and Glyndwr linger in the land.

The greatest modern development of the county has undoubtedly been Llandrindod, which at a bound has become one of the foremost of British spas, and incidentally a much-needed meeting-place of North, Mid, and South Wales for national and provincial purposes.

## 3. Size. Shape. Boundaries. Detached Portions.

Having seen how Radnor came to be formed into a county, and examined its general characteristics as a shire-unit, we must now turn to the map to note its size and boundaries.

Of the Welsh counties, Radnor is the tenth in size, Anglesey and Flint alone being less in area. Carmarthenshire, the largest, is about twice its size. The English county which most closely approximates it is Bedford, while Huntingdon, Middlesex, and Rutland alone possess smaller acreages.

Its length from north to south is 26 miles, and its breadth from east to west measures 29 miles. Its widest part is in the latitude of Rhayader, and its greatest length

extends from Glasbury in the south to the Kerry Hills in
the north.   In shape it bears a remarkably close resemblance
to Africa.

The lines of demarcation between our counties
in many cases follow natural boundaries, but quite as
often are purely artificial.   Sometimes it is difficult to
see why shire boundaries take unlikely courses.   In the
case of Radnor it is not so difficult of explanation, for the
boundary of the county is merely the outside boundary of
the different lordships from which it was made in 1536.
In other words, the line was the compromise of the
relative strength of the opposing lordships at the time.

A large river is from a defensive point of view a very
good boundary; hence the number of Radnor lordships
that rested on the Wye.   This river, from its confluence
with the Elan as far as Rhydspence on the Hereford
border, forms almost the whole of the southern boundary
of Radnorshire, dividing it from Brecon and Hereford.
A little beyond Rhydspence the boundary line turns
northward, and keeps the same general direction in a
jagged line as far as Knighton, a deviation to the west
giving the town of Kington to Hereford, and another to
the east adding Presteign to Radnor.   This section cuts
the valleys of the Arrow and Lug almost at right angles,
and also intersects Offa's Dyke, near Lower Harpton.
And although the general courses of both the ancient and
modern boundary lines are substantially the same, from
south to north, the Dyke is the more easterly of the
two to the south of this point, and the more westerly
north of it.

After striking Shropshire near Knighton, the Radnor boundary makes a sharp turn to the north-west, following the course of the Teme for about 15 miles, after which it ascends the highlands to meet Montgomeryshire under the shadow of the Kerry Hills, when it turns westward, and follows the watershed between Severn and Wye until it touches Cardiganshire in the solitary wilds drained

**The Wye at Hay**
(*Radnor, Brecon, and Hereford Boundary*)

by the Talog, Gwngy, and Nantfigen. It follows the last-named stream to its confluence with the Claerwen, the Claerwen until it meets the Elan, and the Elan till it meets the Wye.

In the Wye section of the boundary, there used to be a little deviation into Brecon from the river bed near Glasbury.  But in the early years of Victoria's reign, this portion of Radnorshire south of the Wye was annexed to Brecon for all purposes.  The Wye now is the dividing line between the two counties, from its junction with the Elan near Rhayader to the town of Hay, and the repairs of Glasbury Bridge, which formerly fell on the county of Radnor alone, are now done at the joint expense of the counties of Radnor and Brecon.

Another matter rectified by the same law was the giving of the detached portion of Hereford—called Lytton Hill—near Cascob, to Radnorshire.  These outliers of certain counties, wholly insulated by other and neighbouring shires, open up a page of very interesting medieval history.  This portion of Hereford in Radnor no doubt belonged at one time to a lord whose chief possessions lay in the former county, and was attached to the hundred he belonged to for the sake of convenience.

The Radnor boundary line measures approximately 95 miles and includes an area computed by the Director-General of Ordnance Survey in 1910 to be 299,521 acres or 468 square miles excluding water.

It is worthy of note that Radnor is not a county easy of local government, as its greatest population lies in the towns and villages around the boundary line, leaving the great central tract of the Radnor Forest very sparsely populated.

## 4. Surface and General Features.

Radnor is the most mountainous part of Fferllys, two-thirds of the whole county being classified as mountain land. Its peaks are not so high nor so numerous as those of the neighbouring county of Brecon, yet the general elevation is greater, owing to the large area of

On Radnor Forest

high moorland which extends from end to end of the county. The average height is well over 1000 feet, and nowhere but in the south-eastern corner is it as low as 300 feet.

The surface of Radnorshire divides itself naturally into (a) Mountain Area, (b) High Moorland, and (c) Valleys.

(*a*)   The collective name for all the mountain tract is Radnor Forest.   This occupies the centre of the county, where Glastwyn and Black Mixon rise to the height of 2186 and 2166 feet respectively.   Clustering around are other mountains approaching them in height, e.g. Bach Hill, The Whimble, and Whinyard Rocks; but with the exception of the first named they do not quite reach the 2000 feet line.

To the west of Rhayader, and overlooking the Elan reservoirs, is a separate group of mountains attaining almost the same height.   Among them are Moel-Fryn ("the bleak hill") 1708, Bwlch-y-Gader ("the hollow of the chair") 1765, and Pen-y-Bwlch ("the head of the hollow") 1633 feet.

At the other end of the county the famous Hergest Ridge of Hereford pushes over the Radnor border to end in Hanter Hill (1361 feet).

(*b*)   Connecting the greatest heights are a number of lesser elevations which sometimes spread out into extensive tablelands, and at others form great ridges enclosing numerous moors, themselves only slightly lower. They are generally designated by the Welsh terms *rhôs* (a swampy moor), *esgair* (ridge), *craig* (rock), *bryn* (hill), *waun* (moor), *carn* (cairn), *moel* (a bleak spot), and *cefn* (back).   Some of these are very noted for their prehistoric remains and for other reasons, as for example Carneddau, near Llanelwedd.

The heights between Painscastle and Glasbury are called Begwns or Beacons, and although they do not approach the Brecon Beacons in height, sufficiently indicate

by their name the uses they were put to when beacon
fires were the readiest method of raising the alarm in time
of war.

(c)   As has been already mentioned, the valleys of
the shire, although small in extent, are very beautiful.
They are extensively wooded and well cultivated.   The
fame of the Wye Valley is world-wide, and the valleys of
the Arrow, Lug, and Teme, if less known, are hardly
less charming, whether considered from the standpoint of
art or that of husbandry.

The valleys of the Elan and Claerwen, in the western
part of the shire, have been almost wholly taken up by
the chain of reservoirs belonging to Birmingham City.
Being at a great elevation they never were so rich as those
of East Radnor, but for the purpose of water-collecting
they are admirable.

## 5.   Rivers and Watershed.   The Smaller Streams.

Radnor possesses two groups of rivers,

(a)   The Wye and its tributaries, running generally
south,

(b)   Some of the tributaries of the Severn, flowing
eastward through Shropshire and Hereford.

The courses of the rivers should be especially noted,
for it is through the gaps here worn by the water-action
of millions of years that the track of the ancient conquerors
led, and the modern road and railway have followed.

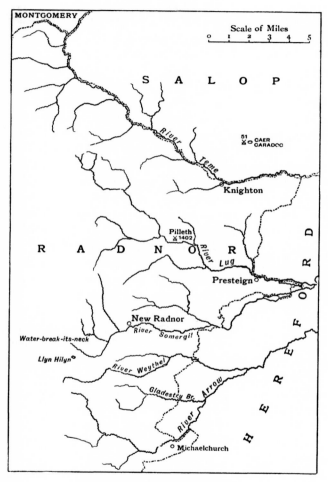

Radnor Rivers flowing East

The Wye, being such an important stream, demands a chapter of its own, and therefore we will first examine the character of the lesser rivers that form the second group.

It may be said of them all that throughout their Radnor courses they are merely limpid mountain streams, excellent for trout, and better still as possible and actual water supplies to the large population toiling in the industrial districts of South Wales and the Midlands. All enter England on the eastern boundary of the county.

Commencing at the south-eastern corner we first strike the Arrow (*garw* or *arw*, i.e. rough). It issues out of a morass on a hill called Waun-y-Gwesta, at the furthest extremity of Colva parish. On its banks are the villages of Glascwm, Newchurch, and Michaelchurch. Near the last-named place it becomes for a few miles the boundary between Radnor and Hereford, before finally entering the latter county to fall into the Lug, about a mile beyond Leominster. Before crossing into Hereford, it skirts the large camp of the Gaer, one of the many ancient fortifications in Radnorshire bearing this name. Its chief tributaries within the county are Glascurnig, Cwmgruffydd Brook, Glasnant, and Cwmgwillen, the last-named issuing from Rhosgoch Pool, in the parish of Bryngwyn.

Next in order comes the Avon, which rises in Colva, and flows through a well-wooded valley, passing through Gladestry, whence it is sometimes called the Gladestry Brook. One of its small feeders formerly supplied the moat of Huntington Castle.

Water-break-its-neck

Making our way still further north we meet the small stream called the Weythel. This also rises in one of the Colva morasses, and after a short course through the village of Weythel, and past the celebrated lime-kilns of Old Radnor, enters the Arrow beyond the confines of the shire.

The Lug Bridge, Presteign

The Somergil claims an importance wholly disproportionate to its size, for many authorities hold that it is owing to its current being lost in underground channels in the Vale of Radnor that the Welsh name of the county —Maesyfed—is derived. Its name also claims some attention, as it is one of the few rivers in Wales which has a distinctly Anglo-Saxon name. It rises in Radnor

Forest, and in its early course forms the wonderful cascade known as Water-break-its-neck, which has a height of 300 feet. It here turns south, and then east, past the town of New Radnor. Its subterranean passage extends two miles, at the end of which it emerges into a large pool of great depth. After continuing on this course for 14 miles, past Downton, Harpton Court, and many

River Teme near Knighton

another ancient seat, it enters Hereford near Lower Harpton.

The Lug is one of the principal Radnor streams flowing east. The name is a corruption of the Welsh "Llugwy" which means "clear water." It rises in a small lake above Llangunllo called Y Ffynnon Aur ("the Golden Well"), and flows past that village and Pilleth, where Rhys the Terrible, Glyndwr's lieutenant, inflicted

a crushing defeat on Edmund Mortimer, in 1402. Thence
it runs through the beautiful Vale of Whitton to Presteign.
After leaving the county-town it soon enters Hereford, at
the foot of Stapleton Castle, to pass ultimately another
fateful battlefield—Mortimer's Cross.

The most northerly of the Radnor streams is the
Teme, which, after rising in the Kerry Hills, flows south-
east through a valley called Dyffryn Tafediad. This
river for the greater part of its Radnor course forms the
boundary between our county and Shropshire. It drains
a very thinly-populated district, which abounds in relics
of bygone days. After dividing the entrenchment of
Crug-y-Byddar into two parts, it skirts the castle of
Cnwclas, and flows through Knighton, past the spot
where Caractacus made his last stand against the Romans.
It enters Hereford near the village of Brampton Bryan.


## 6.  The Wye—(a) **Main Stream.**

The Wye is wholly a Radnorshire river for about
10 miles only. For 34 miles more it divides the county
from Breconshire, and for another 10 miles from Herefordshire.

It has already had a course of 30 miles before reaching
the county at all, so it is a considerable stream when it
enters the border shire, which it does at a place called
Safarn-y-Coed, in the highlands to the north-west of
St Harmon, where the Moel-Fryn rocks stand in imposing
array beside it.   From the wild country on both banks it

receives many a mountain torrent, so that when it reaches
Rhayader (i.e. "the town of the cataract") it has a strong
current. At the lower end of this town, it meets with a

**The Upper Wye and its Radnorshire Tributaries**

rocky obstruction, which has caused the stream to scoop
out great cavities in its bed, forming deep, dark, pools like
Llyn Defaid. As the salmon found great difficulty in
ascending beyond these obstructions, a salmon ladder has

been made below Rhayader bridge to aid them in their
annual upward migration.

A short distance south of the town the river strikes
Breconshire, where Craig-Cnwch on the Radnor side,
and Craig-y-Foel opposite, stand facing each other over its
bed. The current is again impeded for some distance
below this point. Indeed for the greater part of its

The Wye at Builth Bridge

Radnor course the Wye is still a mountain stream. It is
only below the great bend of Llyswen that it assumes the
character of a lowland river. But beautiful as are its
lower reaches, it is doubtful if the upper stream is not of
equal loveliness.

Passing Doldowlod, the Wye flows through a wooded
defile on its way towards Builth, receiving near Newbridge

the Ithon on the left, and near Builth the Irfon on the
right.    Opposite Newbridge (where a still newer bridge
was erected in 1911) the Llysdinam woods clothe
the Brecon bluffs with great beauty.    As the river
approaches nearer to Builth, its bed is very disturbed.
This part—the Builth Rocks, so well known to salmon
anglers—has another series of smoothed-out cavities, where
whirling pebbles have, in the floods of countless years,
formed such dangerous features as Hell Hole, and Aber
Pool.

At Builth the Wye is crossed by a famous bridge of
six arches, towards which all the Brecon and Radnor
roads of the district converge.    Just above this bridge,
owing to the occurrence of fairly level ground, the river
is navigable for some distance for small pleasure boats.
This is the only part of the river touching Radnor ground
that can be so utilised.

No patriotic Welshman can ponder over the history
of this district of Builth and the Wye without feeling a
throb of sympathy for the heroic Llewelyn ap Gruffydd
in his struggle against hopeless odds for the maintenance
of Welsh independence.    But this will be dealt with in a
later chapter.

The narrow valley of the Wye for six miles below
Builth is a good example of the kind of gorge which on a
larger scale is known in America as a cañon.    The sides
of it are well wooded, the heights of Esgoig and Alltmawr
being, in particular, beautifully clothed.    The rocks of
Aberedw are a prominent landmark of this part of the
river, presenting the most striking instance of the bold

General View of the Wye Valley with Builth and Llanelwedd

outcrops of the Silurian system, which rise terrace upon terrace to a height of 700 or 800 feet.

At Erwood, which means Y Rhyd (the ford), there is now a bridge, but the name was evidently given when the "*rhyd*" was the only means of crossing. Opposite this village on the Radnor side is the Garth Hill, from which one of the best views of the Upper Wye may be

Horseshoe Bend of the Wye, Boughrood

enjoyed. The mountains below this point approach the river, and the woods press on the rocky banks.

As we approach Boughrood, now pronounced Bock-rood (a corruption of *Bach rhyd*, "the little ford"), the valley widens out considerably, and the river begins to partake of a lowland character.

Below Boughrood, the river makes a striking horse-shoe bend, which is considered by some to be, in point of beauty, a rival to the better-known "link" near the Windcliff. With a perceptible retarding of its current, it passes ancient Llyswen, and many a stately Tudor and Georgian mansion standing back from the stream.

Glasbury, nestling in a circle of orchards, is a charming district, where the soil gives ample repayment for the culture bestowed on it. Here the bed of the Wye has been diverted, some say, artificially. There is still a chain of pools called *Hên-wy* (the old Wye) to mark the original course of the stream.

Leaving Y Clâs, i.e. the town of the cloisters, as the Welsh call Glasbury, the majestic stream meanders towards Hay, the Norman "enclosure" which long formed the principal gateway into Mid-Wales from the southern Marches. Here, on the Brecon side, is the delightful Bailey Walk, and below it is the Steeple Pool, where the old bells of St Mary's Church are popularly supposed to be lying. The river scenery of this district has been immortalised by David Cox, the great water-colour artist, who did much to enhance the fame of this most beautiful of rivers.

Beyond the town the Brecon boundary falls away from the river, but Radnor continues to rest on it for a further 10 miles, with Hereford on the eastern bank. Llowes and Clyro, both on the left bank, are beautiful districts containing Boatside, Monachty ("the monk's house"), Court-Evan-Gwynne, and Clyro Court, houses well known to artists and antiquaries of many a shire. Opposite Clyro, the river flows in very deep pools, locally called "The Salmon Holes." As at Glasbury, the Wye at this village has been to a small extent artificially diverted for local improvements.

On the right bank of the river at this point stand the ruins of Clifford Castle, with its reminiscences of Fair Rosamond, the unhappy favourite of Henry II. Whitney, on the same side, is a typical Herefordshire village, encircled by orchards and well-cultivated fields.

As the Hereford boundary here crosses to the left bank of the river, the Radnorshire section of the Wye ends at this place near one of the most ancient and quaint inns in the whole country side—Rhydspence—noted for its porch and oak carving.

The stream itself here bids a temporary adieu to the land of the hills, to return after awhile to Nether Went, before finally falling into the Severn Sea, below Chepstow.

## 7. The Wye—(*b*) Tributaries.

Most of the right-bank tributaries of the Upper Wye belong to Brecon, but the Elan, with its feeder, the Claerwen, are Radnor rivers. The left-bank tributaries on the other hand are nearly all of our county.

From north to south they occur in order of sequence thus :—

(*a*) Marteg.

(*b*) Ithon—feeders Clywedog, Cwmaron, and Dulas.

(*c*) Edwy or Edw—feeders Colwyn, Busnant, Lathury, and Rhulen.

(*d*)  Bachwy or Bach-howey.

(*e*)  Wenwy.

The Elan rises in a morass between Rhayader and Cwmystwyth, on the Cardigan border.  The scenery of its narrow vale is strikingly wild.  It was this that appealed so strongly to Shelley, who came here to live before his sojourn in North Wales, but his restless nature found no

The Marteg near St Harmon's

abiding place anywhere, and he soon sought fresh faces and new surroundings.  The houses of Nantgwyllt and Cwm Elan (see p. 138), both connected with his stay in Radnor, now lie fathoms deep beneath the waters of the Birmingham reservoirs, which have practically taken up the whole of the lower course of the Elan and its tributary, the Claerwen.  Their surplus water flows into the Wye, as we have already seen, about a mile below Rhayader town.

(a) The Marteg, i.e. "the handsome river," rises near the Montgomery boundary, and has a mountainous course throughout. In its flow it passes the ancient cell of St Garmon, and joins the Wye at Nannerth, west of Rhayader.

(b) The Ithon, the only river whose course lies wholly in Radnor, rises in Llanbadarn-Fynydd in the north of the county. It has a total length of 30 miles, its numerous meanderings accounting for such a considerable length in so small a county. It passes through several villages of note, mainly interesting on account of the relics of antiquity found in their vicinity. For instance, this stream formed one of the fortifying lines of the Roman station of Magos. It has many windings, too, around Cefnllys Castle.

Itself a tributary of the Wye, the Ithon receives the waters of several well-known lesser streams. The Clywedog passes the ruins of Abbey-Cwm-Hir, where it is conjectured the headless body of Llewelyn the Last lies buried. The Cwmaron has given its name to Cwmaron Castle, now, like almost all the Radnor Norman fortresses, nothing but a site and a mound of débris.

The Dulas is another feeder of the Ithon which must not be confounded with the Dulas which enters the Wye itself at the town of Hay and divides Wales from England at that place. There are several rivers of this name. Apart from the two here mentioned there is one in the Vale of Neath, another a tributary of the Tâf (Carmarthenshire), yet another a tributary of the Llwchwr, and at least one in North Wales, not to speak of Scotland, where a stream of analogous form has given the world the classic name " Douglas."

The course of the Ithon, which ends one and a half miles below Newbridge, is said to run through a more extensive portion of enclosed land than any in the county.

Valley of the Ithon

Lewys Glyn Cothi, the Lancastrian poet of the time of the Wars of the Roses, wrote a fine poem on the beauty of this vale.

(c)   The Edw or Edwy rises in the *rhôs* of Llandegley, and on its way toward the Wye receives the Colwyn, Busnant, Lathury, and Rhulen.   For the first half of its course it passes through swampy ground characteristic of

a Welsh *rhôs*, but in its lower portion, viz. from Colwyn
Castle to Aberedw, its banks are exceedingly fertile.   Near
its termination it has been compelled to force a passage
through a magnificent pile of rocks.   Near its confluence
with the Wye are Llewelyn's Cave and Aberedw Castle,
both closely connected with the last days of Llewelyn
ap Gruffydd.   It was of the Edwy that Drayton in his
*Polyolbion* prettily said that "it bears the message of the
wood nymphs from far Radnorian forests to the Wye."

(*d*)  The Bachwy or Bach-howey rises in the red
*rhôs* of Painscastle, and flows due west to meet the parent
stream.   In its lower course it has carved out a deep and
wild glen for itself which from its interests has from time
to time been visited by several eminent geologists and
naturalists.   Here, near the Craig Pwll Du waterfall, it
was that Sir Roderick Murchison made many geological
discoveries concerning the Silurian System.   It is here, as
well, that some of the rarest of plants are to be found.

The corruption of "Bachwy" to the English form
"Matchway" is typical of a process very common in
Radnorshire, where the place-names are preponderatingly
Welsh and the vulgar tongue almost wholly English.

(*e*)  The Wenwy is a small stream which flows through
a pretty little vale ending at Clyro, opposite the more famous
Vale of Cusop on the south side of the Wye.

In dealing with the streams of this county we cannot
but notice the numerous variations of the Celtic word
"wy," signifying water.

We have had Wy (the Wye) itself,

Llug-wy = the clear Wye,

Bach-wy = the little Wye,
Hen-wy = the old Wye,
Lathur-wy = the smooth Wye,
Arrow(y) = the rough Wye,
Ed-wy = the meandering Wye,
And lastly the Wen-wy = the white Wye.

## 8.  Lakes and Reservoirs.

From the fact that Radnor is wholly an upland
county one would expect to find that, like Carnarvon
and Merioneth, it has a considerable number of natural
lakes.  But it is not so.   Llynbychllyn, its largest sheet
of water, cannot be compared to the Merionethshire
Bala or even the Breconshire Safaddan in point of size.
In character also it is more like a lowland mere than a
mountain tarn, its shores being reedy and marshy.

But, like every other outstanding natural feature of
a superstitious country, Llynbychllyn takes a prominent
place in the folklore and tradition of the district.
Geologists of the twentieth century say that this lake
was formed by a diversion of a stream through the accu-
mulation of glacier débris, and that there are certain signs
that it was originally larger in extent than at present.
But a more romantic story of its origin is given by
Giraldus Cambrensis in his *Itinerary*.   He says, "It came
to pass in the province of Elvenia, which is separated
from Hay by the river Wye, on the night in which
Henry the First expired, that two pools of no small extent,

the one natural and the other artificial, burst their bounds. The latter by its precipitate course down the declivities emptied itself, but the former with its fish and contents obtained a situation in a valley about two miles distant."

Leland in 1540 notices this lake during his tour and says, "There is a *lline* in low Elwel within a mile of Payne's Castel by the church called Llanpeder. The

Llyn Gwyn

Lline is called Boug lline and is of no great quantity but is plentiful of pike and perch and eels."

Llanillin Lake, near Blaen Edw, parish of Llanfi-hangel-Nant-Melan, is a pool of about a mile in cir-cumference, situated in an elevated mountain valley. Horse-racing often used to take place round it, the edge of the lake being the course. Some are of opinion that this lake is of artificial origin.

Llyn Gwyn, to the south-east of Rhayader, is a pretty sheet of water.   It lies on Rhosfa Common, about the centre of the parish of Nantmel, and is of about the same size as Llanillin.   Along one side of it is a semi-circular elevation which traditionally marks the spot where an important town used to be.   The immediate neighbourhood of this pool is well-wooded.   On that account

Pencerrig Lake

chiefly Malkin said it was " the only picturesque lake of Radnor."

Scattered among the highlands of the county are other pools of less size which are but names to any outside their immediate locality, but which play a more important part when considered in connection with local folklore and topography.   Such are the Gaer pool in the mountains of

Llanddewi-Ystrad-Enny situated near to no less than three ancient camps, and the Penpergwm lakes near Glasbury, which are entirely due to glacial drift and are hollows in the drift itself. Pencerrig Lake near Llanelwedd, and Hendwell Pool, formed by the Somergil brook near Old Radnor, are others worthy of mention.

At Cefnpawl in the parish of Abbey-Cwm-Hir there used to be a remarkably large fishpond which supplied the monks with fish.

But all the ancient lakes of Radnor—natural and artificial—pale into insignificance when compared with the Elan series of lakes recently formed in the north-west of the county. Both in size and sublimity of surroundings they are beyond anything that South Wales can show, and so well constructed are they, that for ages to come they will doubtless form one of the most striking features of this border shire.

Some one has said that in Wales one can always hear the sound of running water, and to no counties in the Principality is the statement more applicable than to Radnor and Brecon. Numerous streams of pure clear water flow from the hillsides in all directions, unpolluted by the refuse from any great industrial undertaking. It is no wonder, then, that many populous towns come to these Welsh hills for their water-supply and that so many important waterworks lie within, or are situated near, the confines of these counties. Here, for their water, come Birmingham, Cardiff, Swansea, and Merthyr, while lesser works supply Neath, Aberdare, Ebbw Vale, and other large towns, and an immense scheme is even under

consideration whereby the Breconshire Irfon will be tapped
for the purpose of supplying London with Welsh water.

The most important of the waterworks are those
constructed by the city of Birmingham in the valleys
of the Elan and the Claerwen. Part, only, of these
works lies within the county, the border-line between

Caban Coch Dam

Breconshire and Radnorshire running along the line
taken by the ancient courses of the streams in this part.
The Caban Coch dam, formed at the junction of the
Claerwen and the Elan, is the first of a series of dams
which have transformed the picturesque Elan valley into
the three huge artificial lakes of the Caban Coch com-
pensation reservoir, and the Penygareg and Craig Goch

supply reservoirs. The first dam stands at an elevation of 820 feet, the Penygareg dam at an altitude of 945 feet, and the Craig Goch dam at 1040 feet, the water surface area of the three reservoirs being 497 acres, 124 acres, and 217 acres respectively. At Caregddu, where the Caban Coch lake shoots out one arm to the north along the Elan valley and another to the south along the valley

Caregddu Reservoir, Elan Valley

of the Claerwen, is a huge submerged dam from which the conduit leads. These lakes contain 11,000 million gallons of water, and yield a daily supply to Birmingham of 27 million gallons.

In the Claerwen valley is the Dolymynach dam, which impounds a mass of water with a surface area of 148 acres. Further dams are to be constructed in the Claerwen valley;

one at Cil Oerwynt at an altitude of 1095 feet will create a reservoir of 269 acres surface area, and one at Nant y Beddau at 1175 feet altitude will impound a lake with 244 acres of surface. Altogether the city of Birmingham has acquired upwards of 40,000 acres of land in the watershed, thereby ensuring that the sources whence the city derives its water shall be absolutely uncontaminated.

The Filtering Beds, Cwm Elan Reservoirs

In all, this stupendous undertaking will cost some £6,000,000.

The aqueduct, from the intake to Birmingham, is some 73½ miles long. For the first 36½ miles it consists of a brick and concrete structure 9 feet in diameter, 13½ miles of which is in tunnel and 23 miles in cut and cover. For the remaining 37 miles the water is conveyed in huge iron pipes.

## 9.  Fisheries.

There has been a time—and not so long ago—when the Wye was considered a salmon-river of the first class.

It no doubt had to supply its share when Edward II, wishing in 1308 to set his troops in motion to wage war

Salmon Ladder, Rhayader Bridge

against Scotland, requisitioned 3000 dried salmon from Wales to victual them.  But the Domesday Book, compiled almost two and a half centuries before that date, mentions the fishermen of Downton, and Giraldus Cambrensis, speaking of Siluria towards the close of the twelfth century, writes, "River fish are plentiful, supplied by the Usk on one side and by the Wye on the other; each of them produces salmon and trout, but the Wye abounds most with the former and the Usk with the latter. The salmon of the Wye are in season during the winter, those of the Usk in summer, but the Wye alone produces the fish called umber" (grayling).

Shakespeare makes Fluellen say, "And there is salmons in both"; and Churchyard, the Elizabethan rhymester of Ludlow, corroborates Giraldus when he says:—

> "A thing to note when sammon failes in Wye
> (And season there goes out, as order is)
> Then still of course in Oske doth sammons lye
> And of good fish in Oske you shall not miss.
> And this seems straunge as doth through Wales appeare
> In some one place are sammons all the yeere
> So fresh, so sweete, so red, so crimp withall
> That man might say, loe, sammon here at call."

But owing to a variety of causes, e.g. a series of bad seasons and the prevalence of pike and other coarse fish, the Wye about 20 years ago greatly suffered in reputation. Since that time, however, it has to some extent retrieved itself. In the winter of 1902–3, for instance, there was a somewhat exceptional quantity of salmon on the spawning "redds," and hence the numerous conflicts with poachers at that time. In 1908, too, many more spring fish were caught than had been the average for a generation, while the restriction of net fishing which has been in operation for the last three years has had a most beneficial result on the river generally.

Salmon Netting at Llanelwedd

The Wye salmon record for the last few years is as follows:—

|  | | | Number killed. | Average weight. |
|------|---|---|------|-----------|
| 1906 | ... | ... | 468 | $17\frac{1}{4}$ lb. |
| 1907 | ... | ... | 1424 | 14 lb. |
| 1908 | ... | ... | 1571 | $17\frac{1}{4}$ lb. |
| 1909 | ... | ... | 1356 | 15 lb. |
| 1910 | ... | ... | 2715 | $14\frac{1}{2}$ lb. |

In 1910 no less than 84 fish scaled 27 lb. and upwards, the heaviest of all being $46\frac{1}{2}$ lb.

A century ago Theophilus Jones, the county historian of Brecon, enumerated the fish of the Wye to be salmon, trout, grayling, pike, perch, dace, loach, gudgeon, eel, lamprey, roach, bullhead, minnow, and shad, besides crayfish. He also makes the remarkable statement that the crayfish or freshwater lobster is found in many brooks running into the Wye, but seldom, if ever, in those which fall into the Usk or Irfon. Many unsuccessful attempts have been made to remove it into the rivers of Glamorgan and Carmarthen, and even into some brooks of the Irfon, which empties itself into the Wye, but in vain.

Since Jones wrote, the Wye has been visited by at least one other species of fish, for it is recorded that on the 23rd of July, 1832, a sturgeon was caught near Boughrood weighing 131 lb. and measuring 7 ft. 6 in. in length.

It cannot be claimed that the main stream of the Wye is a first-class trout river. One of its Brecon tributaries—the Llyfni—is unfortunately connected with Lake Safaddan, which is a prolific breeding-place of the pike, and these, finding their way into the main stream, destroy more trout than all the nets and lines put together.

The presence of that very local fish, the allice shad, is also a certain signal for the emigration of the trout, but occasionally fine specimens are still caught, one landed at Llyn Defaid near Rhayader Bridge in 1904 weighing $5\frac{1}{2}$ lb.

Of the Wye tributaries the Edw once had a great reputation for the number and quality of its trout, as had the Ithon in a less degree. But the latter has fallen on evil days, as its waters swarm with chub. The former, too, has been sorely depleted and cannot be compared to what it was half a century ago.

The Arrow, Teme, and Lug, all outside the Wye system, still contain an abundance of trout and grayling. The first-named is also credited with containing the crayfish, whose vagaries in the matter of habitat we have already noticed.

Of the natural lakes Llynbychllyn, like Safaddan of Breconshire, teems with pike and perch. Leland in 1540 noticed the same fact. It appears to bear out the remarkable assertion in Natural History that pike, although abnormally voracious, will not attack the much smaller perch. Llynillyn Pool abounds with carp and large eels, while Hendwell Pool contains also lake trout.

The Elan reservoirs have been well stocked with trout, and bid fair to become one of the most noted fishing districts south of the Tweed.

## 10. Geology and Soil.

By Geology we mean the study of the rocks, and we must at the outset explain that the term *rock* is used by the geologist without any reference to the hardness or compactness of the material to which the name is applied; thus he speaks of loose sand as a rock equally with a hard substance like granite.

Rocks are of two kinds, (1) those laid down mostly under water, (2) those due to the action of fire.

The first kind may be compared to sheets of paper one over the other. These sheets are called *beds*, and such beds are usually formed of sand (often containing pebbles), mud or clay, and limestone, or mixtures of these materials. They are laid down as flat or nearly flat sheets, but may afterwards be tilted as the result of movement of the earth's crust, just as we may tilt sheets of paper, folding them into arches and troughs, by pressing them at either end. Again, we may find the tops of the folds so produced worn away as the result of the erosive action of rivers, glaciers, and sea-waves upon them, as we might cut off the tops of the folds of the paper with a pair of shears. This has happened with the ancient beds forming parts of the earth's crust, and we therefore often find them tilted, with the upper parts removed.

The other kinds of rocks are known as igneous rocks, and have been melted under the action of heat and become solid on cooling. When in the molten state they have been poured out at the surface as the lava of volcanoes, or have been forced into other rocks and cooled in the cracks and other places of weakness. Much material is also thrown out of volcanoes as volcanic ash and dust, and is piled up on the sides of the volcano. Such ashy material may be arranged in beds, so that it partakes to some extent of the qualities of the two great rock groups.

The relations of such beds are of great importance to geologists, for by means of these beds we can classify the rocks according to age. If we take two sheets of paper, and lay one on the top of the other on a table, the upper one has been laid down after the other. Similarly with two beds, the upper is also the newer, and the newer will remain on the top after earth-movements, save in very exceptional cases which need not be regarded here, and for general purposes we may look upon any bed or set of beds resting on any other in our own country as being the newer bed or set.

The movements which affect beds may occur at different times. One set of beds may be laid down flat, then thrown into folds by movement, the tops of the beds worn off, and another set of beds laid down upon the worn surface of the older beds, the edges of which will abut against the oldest of the new set of flatly deposited beds, which latter may in turn undergo disturbance and renewal of their upper portions.

Again, after the formation of the beds many changes may occur in them. They may become hardened, pebble-beds being changed into conglomerates, sands into sandstones, muds and clays into mudstones and shales, soft deposits of lime into limestone, and loose volcanic ashes into exceedingly hard rocks. They may also become cracked, and the cracks are often very regular, running in two directions at right angles one to the other. Such cracks are known as *joints*, and the joints are very important in affecting the physical geography of a district. Then, as the result of great pressure applied sideways, the rocks may be so changed that they can be split into thin slabs, which usually, though not necessarily, split along planes standing at high angles to the horizontal. Rocks affected in this way are known as *slates*.

If we could flatten out all the beds of England, and arrange them one over the other and bore a shaft through them, we should see them on the sides of the shaft, the newest appearing at the top and the oldest at the bottom, as in the annexed table. Such a shaft would have a depth of between 10,000 and 20,000 feet. The strata beds are divided into three great groups called Primary or Palaeozoic, Secondary or Mesozoic, and Tertiary or Cainozoic, and the lowest of the Primary rocks are the oldest rocks of Britain, which form as it were the foundation stones on which the other rocks rest. These may be spoken of as the Pre-Cambrian rocks. The three great groups are divided into minor divisions known as systems. The names of these systems are arranged in order in the table, and the rocks which are found in Radnorshire are also stated.

With these preliminary remarks we may now proceed to a brief account of the geology of the county.

In the east of England the rocks of the Secondary and Tertiary periods are the only ones that reach the

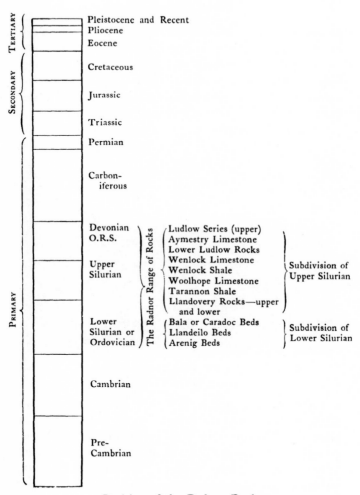

Position of the Radnor Rocks

surface to form the landscape, whereas in Wales the older rocks are much the more common. It is owing to the greater hardness of these that Wales is so rugged and hilly. The words Cambrian, Ordovician, and Silurian are all names of well-known series of rocks which have been so called from their prevalence in the Principality.

We shall probably best understand the geology of

**Aberedw Rocks**
(*Outcrop of Silurian Beds.*)

South Wales (and Radnor as a particular part of it) if we compare this half of the Principality to an immense dish with the Silurian, Old Red Sandstone, Carboniferous Limestone, Millstone Grit, and the Coal Measures placed in it in regular order from the bottom upwards. Its greater length is from east to west, and Radnor forms, as it were, the middle portion of its northern lip, where the rocks crop out one after the other according to their varying depth.

The Coal Measures, Millstone Grit, and Carboniferous Limestone have all cropped out in southern Breconshire, Pencerrig-calch in the Black Mountains being the last outlier of the Limestone nearest the Radnor border.

The Old Red Sandstone or Devonian Rocks cross over from Hereford to central Brecon and cut a section of south Radnor on their way, the rocks around Glasbury being about the centre of the belt. There are portions

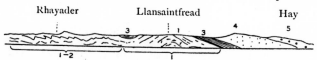

Rhayader          Llansaintfread                    Hay

**Diagrammatic Section from the County Boundary N.W. of Rhayader to Hay**

(*Length of Section about 30 miles*)

1 Lower Silurian rocks with Igneous rocks.
1—2 Highly poised Lower Silurian and Llandovery rocks of the west
3 Wenlock          4 Ludlow          5 Old Red Sandstone

of the same beds further north which are wholly insulated in the next—the Silurian—belt. They are very plainly seen in the neighbourhood of Norton, between Knighton and Presteign. But much the greater part of Radnor belongs to the Silurian system, Upper and Lower. It was this county that Sir Roderick Murchison—the greatest authority on the Silurian beds—made his chief field of study.

The Lower Silurian or Ordovician Rocks are very old, and tell of a period when the face of the earth was inhabited by nothing higher in the scale of life than chambered shells.

D. R.

4

By a reference to the geological map at the end of the book it will be seen that the oldest rocks of Radnor crop out in the north, and that from the Ordovician beds to the Old Red Sandstone, i.e. from north to south, there is a series of steps, the majority of them belonging to the Silurian System. The border line between each is not at all

Erratic Boulder on Carneddau

regular, as portions of one bed often enter far into the field of the succeeding one.

It must not be imagined that even the same systems have the same character throughout their courses. In one locality a particular feature of the Silurian beds, e.g. sandstone, may be emphasised, while in another neighbourhood the outstanding characteristic will be limestone or shale. This accounts for so many *local* geological names being given to the rock prominent in certain places. For instance we have Woolhope Limestone beds near Old Radnor, the Wenlock rocks, the Tarannon shales, the Llandovery rocks, and the Llandeilo shale in other parts of the county. The relative positions of these locally-named beds, as shown in the section of the Silurian system, can be seen by further reference to the table on page 47.

Bursting through all the strata of Silurian and Old Red Sandstone we find at different points of the shire the volcanic or trap rocks. Such is the case at Carneddau in the west, Carregwiber in the centre, and Hanter Worsel and Stanner in the east. Carneddau in particular is a very important geological district, its ice-markings, perched rocks, and moraines being very numerous.

These igneous rocks are of the greatest importance to the social life and business of the county, for they are the origin of the medicinal springs. The igneous matter coming in contact with the schist has produced much sulphuret of iron, the decomposition of which gives rise to the various sulphur, saline, magnesium, and chalybeate springs of Llandrindod, Llandegla and other places. These rocks are continued into Breconshire, where the Builth, Llangammarch, and Llanwrtyd Wells are situated.

It would be natural to surmise that Radnor owing to its variety of rock foundations has also a corresponding variety of soils, and we find this to be the case.

One old writer had noticed 120 years ago that the soil in the Builth neighbourhood was remarkably argillaceous or clayey, and that the water therefore was prevented from sinking sufficiently deep, and was held upon the surface until it soured. As the Wye valley is descended, however, there is a great improvement in the quality of the soil. It loses its injurious tenacity and admits a portion of loam and sand.

Another authority states that the soil of Glasbury on the banks of the Wye is the pride of the country side. Broadly speaking the best portions of the county in this respect are those where the Old Red Sandstone rocks predominate. These are the Glasbury, Clyro, Painscastle, Norton, and Presteign districts. A notable exception is the valley of the Teme, where the land around Knighton, though not of the red-earth variety, is of very good quality. The limestone valleys, though not barren, are at a disadvantage owing to the extreme thinness of the soil, the rocks in many neighbourhoods being struck immediately beneath the turf.

The report of John Clarke who wrote in 1794, although in error as regards the area of the county, remains true to-day as to the proportion of the various characters of soil in the county. He estimated Radnor to contain 326,400 acres, and of these he classified:

     86,400 acres as tillage land,

     40,000 as meadows, pastures and woods,

and   200,000 as common mountain land,

     326,400

which gives one acre in four as consisting of good land.

## 11. Natural History.

Closely dependent on the geological formation of a county is the record of its flora and fauna.

It is well known that our islands at a remote period became entirely submerged. This of course would destroy the plant and animal life then existing. But quite as certain is the fact that since that time the British Islands have been connected with the Continent, and by this means were stocked anew with plants and animals.

But it takes time to stock a land, and consequently the nearest districts may become abundant in species when the more remote have not been reached. When separation occurred later, as it did, and the North Sea and the Channel interposed their barriers, only comparatively few of the continental species had spread so far west as Ireland before that island, in its turn, had become cut off from England. Hence it comes about that the Continent has more species than any part of England, while the south and east of our land are richer than its more remote northern and western parts, and these again richer than Ireland. There are, however, many other factors which affect the number of species in any given locality.

We have noticed in the last chapter that Radnor has quite a variety of soils. We have also seen that while it possesses a number of sheltered valleys it has, in addition, a great expanse of upland meadows as well as a portion of a wholly mountainous character. These facts supply the conditions for a great diversity of flora and such diversity is found to exist.

The place-names of the county show that some of the larger fauna used to be numerous. There is Bleddfa ("the abode of wolves") in the north and the Wolfpits in the east. Indeed, there is a tradition that the last wolf of South Wales was killed in early Tudor times at Cregrina, between Builth and Glascwm. Near Glascwm is Cwm-Twrch ("the valley of the boar") and near Glasbury Cil-Twrch ("the boar's hiding-place").

The marten, although not quite extinct, must be almost so. The last recorded in the county was in the year 1898. *Bela* is the marten's Welsh name. The word exists in such place-names as Cwmbele near Painscastle and Llanbele (Gladestry).

The wild cat has apparently only recently become extinct; the polecat, though not quite killed out, is fast disappearing. Stoats and weasels still abound; foxes, badgers, and otters are also numerous, although the otter's skin is not now an article of the commerce of Builth as it was at the time of the Beaufort Progress in 1684.

Even as late as Leland's day the county abounded with deer. There is a tradition of a deer-park seven miles in circumference at Abbey-Cwm-Hir, and some old ruins are still pointed out as the site of two large deerhouses in connection with the Abbey. Ffos-yr-hyddod ("the ditch of the deer") is mentioned in an old lease of the Grange of Cwmdauddwr dated 1624, and the county has numerous other place-names making reference to these animals.

The county has at least five kinds of bats, viz. the Pipistrelle, the Whiskered, the Noctule, the Long-eared and the Barbastelle. The last was seen for the first time

in 1904 in the porch of Llanelwedd Church. Previous
to that it had not been recorded west of Worcester. It
is believed that Daubenton's bat also exists in the county.

Radnor is now remarkably destitute of snakes, even
the common grass-snake being but seldom seen. But
the name Carregwiber ("the rock of the viper") near
Llandrindod, implies a once different state of things.
The Wellfield rocks near Llanelwedd also were known
to be infested with adders about half a century ago.
Lizards and slow-worms are still plentiful.

Our county has a great variety of birds, especially of
those that are approaching extinction in Great Britain.
The kite, the buzzard, the raven, and the carrion crow
for instance still make their homes in the wilds of the
Matchway and Cwm Elan, while other birds, such as the
peregrine falcon, stock-dove, nuthatch, greater and lesser
spotted woodpecker, are perceptibly increasing in numbers
after having once been in danger of total extinction. The
crossbills are very erratic in their visits, being in some
years very plentiful, while in others they forsake the
county almost entirely.

The following uncommon birds are known to have
bred within the county :—pied flycatcher, spotted crake,
kite, buzzard, raven, hobby, little owl, long-eared owl,
dipper, wryneck, woodlark, quail, curlew, nightjar, lesser
and greater spotted woodpeckers ; while the following
have also been recorded as occurring : hoopoe, grey
phalarope, hen-harrier, pine grosbeak, pomatorhine skua,
green sandpiper, and (a rarity in these western lands) the
garden-warbler.

Radnor is poor in its number of butterflies but has a great number of moths, many rare specimens of which are found within the county.　The Cusop Dingle, situated mainly in Hereford, but right on the Radnor boundary, is one of the most famous entomological districts in Great Britain.　In it are found many rare insects, and one (*Platypeza hirticeps*) found there in 1899 was new to science.

Timber Felling in Radnorshire

Radnorshire can show some fine trees.　Its excellent timber was noticed by writers of two or three centuries ago, notably by Dinely in 1684 and Malkin at the close of the eighteenth century.　John Clarke in 1794 said : "Few if any mountainous districts of this island are so well adapted by nature to the propagation of timber." To-day, as in Clarke's day, "oak and ash are the great favourites of this soil," the oaks of Pencerrig, Abbey-Cwm-Hir, and Harpton being specially famous. It was at the first-named place that the oak forming the keelson of the ill-fated "Royal George" was grown. The large oaks of Abbey-Cwm-Hir were doubtless planted by the monks before the dissolution of the monasteries. The

**The Stanner Rocks**
(*Formed by Volcanic Action.*)

valleys generally may be said to be well wooded with larch, ash, and fir plantations, especially the Wye valley below Builth. Perhaps the most interesting trees of all are the yews which add so much dignity to Glasbury, Old Radnor, and other places. When the bow was the chief weapon of our land the supply of yews was a national question.

Although Radnor has such diverse types of plants it cannot be said to have a large proportion of the 2047 species of flowering plants and ferns which are known to be indigenous to our country, though it has some of the rarest kinds. The Stanner Rocks are so rich in this respect that they have been locally known as the "Devil's Garden." The glen of Craig-Pwll-Du on the Bachwy is a special resort of botanists for the many rare species growing there.

The Radnor Forest bears the Welsh poppy (*Meconopsis cambrica*)—found only in ten counties of Great Britain—Wyeside has toothwort, Llanelwedd chives, Cwm Elan sundew and butterwort, Llynbychllyn the flowering rush and *Osmunda regalis*, and Presteign the ivyleaved toadflax, curved stonecrop, henbane, and *Moenchia erecta*, which is uncommon.

## 12.   Climate and Rainfall.

The climate of a country or district is, briefly, the average weather of that country or district, and it depends upon various factors, all mutually interacting, upon the latitude, the temperature, the direction and strength of the winds, the rainfall, the character of the soil, and the proximity of the district to the sea.

The differences in the climates of the world depend mainly upon latitude, but a scarcely less important factor is this proximity to the sea. Along any great climatic zone there will be found variations in proportion to this proximity, the extremes being "continental" climates in the centres of continents far from the oceans, and "insular" climates in small tracts surrounded by sea. Continental climates show great differences in seasonal temperatures, the winters tending to be unusually cold and the summers unusually warm, while the climate of insular tracts is characterised by equableness and also by greater dampness. Great Britain possesses, by reason of its position, a temperate insular climate, but its average annual temperature is much higher than could be expected from its latitude. The prevalent south-westerly winds cause a drift of the surface-waters of the Atlantic towards our shores, and this warm-water current, which we know as the Gulf Stream, is the chief cause of the mildness of our winters.

Most of our weather comes to us from the Atlantic. It would be impossible here within the limits of a short chapter to discuss fully the causes which affect or control weather changes. It must suffice to say that the conditions are in the main either cyclonic or anticyclonic, which terms may be best explained, perhaps, by comparing the air currents to a stream of water. In a stream a chain of eddies may often be seen fringing the more steadily-moving central water. Regarding the general north-easterly moving air from the Atlantic as such a stream, a chain of eddies may be developed in a belt parallel with its general direction. This belt of eddies, or cyclones as they are termed, tends to shift its position, sometimes passing over our islands, sometimes to the north or south of them, and it is to this shifting that most of our weather changes are due. Cyclonic conditions are associated with a greater or less amount of atmospheric disturbance; anticyclonic with calms.

The prevalent Atlantic winds largely affect our island in another way, namely in its rainfall. The air, heavily laden with moisture from its passage over the ocean, meets with elevated land-tracts directly it reaches our shores—the moorland of Devon and Cornwall, the Welsh mountains, or the fells of Cumberland and Westmorland —and blowing up the rising land-surface, parts with this moisture as rain. To how great an extent this occurs is best seen by reference to the accompanying map of the annual rainfall of England, where it will at once be noticed that the heaviest fall is in the west, and that it decreases with remarkable regularity until the least fall is reached on our eastern shores. Thus in 1908, the maximum rainfall for the year occurred at Llyn-Llydaw in the Snowdon district, where 237 inches of rain fell; and the lowest was at Bourne in Lincolnshire, with a record of about 15 inches. These western highlands, therefore, may not inaptly be compared to an umbrella, sheltering the country further eastward from the rain.

The above causes, then, are those mainly concerned in influencing the weather, but there are other and more local factors which often affect greatly the climate of a place, such, for example, as configuration, position, and soil. The shelter of a range of hills, a southern aspect, a sandy soil, will thus produce conditions which may differ greatly from those of a place—perhaps at no great distance—situated on a wind-swept northern slope with a cold clay soil.

Radnor being an inland county is not tempered by the sea to the extent that Pembroke, with its long sea-coast, is. Again, on account of its being at a greater altitude than Glamorgan, it cannot be expected to be as warm as that county. On the other hand, the fact that many of the Radnor districts are sheltered by high hills tends to raise the general temperature of the valleys, while abundance of trees and other vegetation also helps. The position of the shire with its mountain wall in the path of the warm westerly winds has an important bearing on its general climate and particularly on the abundance of its annual rainfall. When the resultant of all these factors, as expressed in degrees of temperature and in number of hours of bright sunshine, is examined, Radnor cannot be considered a sunny county.

The Llangammarch station (just outside the Radnor boundary) for the month of April, 1910, registered 112 hours of bright sunshine, whereas Cardiff for the same month received 141 hours, Rhyl 144 hours, Pembroke 150 hours, Bournemouth 153 hours, and Lowestoft 189 hours. During the month of November of the same year the difference was still more noticeable, for while Lowestoft received 107 hours, Bournemouth 100 hours, Cardiff 91 hours, Pembroke 78 hours, and Rhyl 75 hours, Llangammarch showed only 59 hours.

The great Archdeacon of Brycheiniog, Gerald the Welshman, had noticed some time about the end of the twelfth century that northern Siluria suffered from a lack of sunshine, but he expresses it in a different and more pleasant manner, for he says, "Being thus sheltered on the

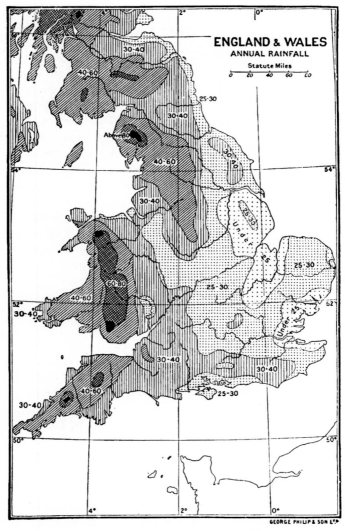

*(The figures give the approximate annual rainfall in inches.)*

south by high mountains the cooler breezes protect this district from the heat of the sun."

No observant visitor to a breezy Radnor moor can fail to notice the direction of its prevalent winds. The trees, large and small, whether they be oak, birch, or mountain ash, all grow somewhat bent to the east and north-east as the result of continuous west and south-west winds. But the great feature of the westerly winds as bearing on the climate of Radnor is the high average of annual rainfall they bestow on it. It is this fact, in conjunction with other favourable conditions, that has made Radnor the water-provider of the midlands.

When we come to look into the Radnor rainfall records we find that some of the figures tell a remarkable tale. While Cardiff over an extended period has shown an average rainfall of 37 ins. and Neath of 47 ins., Nantgwyllt in Cwm Elan has averaged 61·74 ins. over three decades. In certain years this high average has been greatly exceeded, for in 1872 it totalled 93·86 ins., and the rainfall of 43·43 ins. in 1892, although well over the English average (33·61 ins.), was considered extremely low for Radnor.

The floods of Fferllys have played an important part even in our national history. It is recorded that it was owing to the inundation of the Severn and its tributaries that Owen Glyndwr failed to join Hotspur in time for the Battle of Shrewsbury (1403). Again in 1483 it was a flood that probably saved the throne of England for Richard III, when the Duke of Buckingham had raised his standard of revolt at Brecon Castle and had commenced

The Wye in Flood

his eastward journey through south Radnor to join his confederates. But the roads and fords became impassable and the whole scheme failed in consequence, and to this day the flood is spoken of as "Buckingham's Deluge."

## 13. People—Race, Settlements, Population.

We have no written record of the history of our land carrying us beyond the Roman invasion in B.C. 55, but we know that Man inhabited it for ages before this date. The art of writing being then unknown, the people of those days could leave us no account of their lives and occupations, and hence we term these times the Prehistoric period. But other things besides books can tell a story, and there has survived from their time a vast quantity of objects (which are daily being revealed by the plough of the farmer or the spade of the antiquary), such as the weapons and domestic implements they used, the huts and tombs and monuments they built, and the bones of the animals they lived on, which enable us to get a fairly accurate idea of the life of those days.

So infinitely remote are the times in which the earliest forerunners of our race flourished, that scientists have not ventured to date either their advent or how long each division in which they have arranged them lasted. It must therefore be understood that these divisions or Ages—of which we are now going to speak—have been adopted for convenience sake rather than with any aim at accuracy.

The periods have been named from the material of which the weapons and implements were at that time fashioned—the Palaeolithic or Old Stone Age; the Neolithic or Later Stone Age; the Bronze Age; and the Iron Age. But just as we find stone axes in use at the present day among savage tribes in remote islands, so it must be remembered the weapons of one material were often in use in the next Age, or possibly even in a later one ; that the Ages, in short, overlapped.

Let us now examine these periods more closely. First, the Palaeolithic or Old Stone Age. Man was now in his most primitive condition. He probably did not till the land or cultivate any kind of plant or keep any domestic animals. He lived on wild plants and roots and such wild animals as he could kill, the reindeer being then abundant in this country. He was largely a cave-dweller and probably used skins exclusively for clothing. He erected no monuments to his dead and built no huts. He could, however, shape flint implements with very great dexterity, though he had as yet not learnt either to grind or polish them. There is still some difference of opinion among authorities, but most agree that, though this may not have been the case in other countries, there was in our own land a vast gap of time between the people of this and the succeeding period. Palaeolithic man, who inhabited either scantily or not at all the parts north of England and made his chief home in the more southern districts, disappeared altogether from the country, which was later re-peopled by Neolithic man.

Neolithic man was in every way in a much more

advanced state of civilisation than his precursor. He tilled the land, bred stock, wore garments, built huts, made rude pottery, and erected remarkable monuments. He had, nevertheless, not yet discovered the use of the metals, and his implements and weapons were still made of stone or bone, though the former were often beautifully shaped and polished.

Between the Later Stone Age and the Bronze Age there was no gap, the one merging imperceptibly into the other. The discovery of the method of smelting the ores of copper and tin, and of mixing them, was doubtless a slow affair, and the bronze weapons must have been ages in supplanting those of stone, for lack of intercommunication at that time presented enormous difficulties to the spread of knowledge. Bronze Age man, in addition to fashioning beautiful weapons and implements, made good pottery, and buried his dead in circular barrows.

In due course of time man learnt how to smelt the ores of iron, and the Age of Bronze passed slowly into the Iron Age, which brings us into the period of written history, for the Romans found the inhabitants of Britain using implements of iron.

We may now pause for a moment to consider who these people were who inhabited our land in these far-off ages. Of Palaeolithic man we can say nothing. His successors, the people of the Later Stone Age, are believed to have been largely of Iberian stock; people, that is, from south-western Europe, who brought with them their knowledge of such primitive arts and crafts as were then discovered. How long they remained in undisturbed possession of our land we do not know, but they were later conquered or driven westward by a very different race of Celtic origin—the Goidels or Gaels, a tall, light-haired people, workers in bronze, whose descendants and language are to be found to-day in many parts of Scotland, Ireland, and the Isle of Man. Another Celtic

Foundation of Goidel Hut, Carneddau

people poured into the country about the fourth century B.C.—the Brythons or Britons, who in turn dispossessed the Gael, at all events so far as England and Wales are concerned. The Brythons were the first users of iron in our country.

The Romans, who first reached our shores in B.C. 55, held the land till about A.D. 410; but in spite of the length of their domination they do not seem to have left much mark on the people. After their departure, treading close on their heels, came the Saxons, Jutes, and Angles. But with these and with the incursions of the Danes and Irish we have left the uncertain region of the Prehistoric Age for the surer ground of History.

We have now arrived at the time of the Anglo-Saxon kingdoms. The nearest of them to Radnor was Mercia, which gradually became strong enough to keep the British within their mountain valleys. Offa's Dyke, which marked the line of compromise after several centuries of fighting, skirts the eastern boundary of the modern Radnor, and beyond this Mercia failed to keep a permanent footing.

But what Offa's descendants failed to do the Saxons of Wessex accomplished, for Harold Godwin certainly established a colony in New Radnor before he was called to the English throne.

The Irishman from the west had also some measure of success here, although not to the same extent as in the neighbouring Brecon, where an Irish prince became the founder of the long line of Brycheiniog chieftains. Still, Cwmgwyddel ("the vale of the Irishman") in Nantmel, and Cytiau'r Gwyddelod ("the Irishmen's huts") in various parts, testify to their presence in those far-off times. By their adoption of the language of the country they were soon lost as a separate entity and the two nations united, with Welsh as their common speech, so that when the Norman came Radnor was apparently as Welsh in character as ever, although in race the people were a mixture, the descendants of the Silures doubtless preponderating, but

with a large infusion of Irish and a weaker infusion of
Roman and Saxon blood.

Offa's Dyke was something more than a political
boundary. It was also a sharp dividing-line between the
English and Welsh languages. For over a thousand years
Welsh was as prevalent within a mile or two west of it
as English was to the east of it.

When the Norman-French crossed into Wales they
freely intermarried with the natives, in spite of numerous
decrees of monarchs and parliaments forbidding them to
do so, and Radnor to-day has many families such as
Rogers, Progers, Jeffrey, Pritchard, Bailey, Venables,
Harris, and Mortimer, to testify to the fact. But they too
fell under the spell of the ancient tongue, and became
in time as Welsh-speaking as the Vaughans, Gwynnes,
Prices, Parrys, and Powells themselves.

It was towards the close of the eighteenth and the
beginning of the nineteenth century that Radnor became
Anglicised. A Shropshire solicitor touring the county in
1744 says of the frequenters of the Disserth feast that
"they played very well, but spoke, as *almost everyone else
did*, in the Welsh tongue." The same thing struck him
at Aberedw, for he says, "Against the tower were some
of the most active lads I have seen playing at fives. Their
language was *entirely Welsh*." Lewis Morris, the ancestor
of the author of the *Epic of Hades*, also said in 1747 that
"in all the Radnor Churches divine service was given
in Welsh alone." But in 1818, Williams, the county
historian, deplored its decadence. And from that time
its neglect has been so great that Radnor commences

the twentieth century as practically wholly an English-speaking county.

This has not come about by any great immigration, for, beyond a few Scots attracted to its sheep-walks, there has been no national invasion of any kind since the Norman period.   Here, as in some parts of Brecon and Monmouth, it is owing to the indifference of the Welshmen themselves that they have become gradually Anglicised.   It is a pleasant fact, however, to record that although the present Radnorshireman has lost the speech of his forebears, he still cherishes his country's great institution—the Eisteddfod—Llandrindod and other places within the county having periodic meetings in connection therewith.

Having seen what the men of Radnor were like in times past, let us now further glance at their descendants of to-day.

There was no regular census in England and Wales before 1801, but the population of Radnor was computed about the middle of the eighteenth century to vary between 17,000 and 18,000.   This number in 1801 had increased to 19,700, whereas in 1911 the number was shown to be 22,589.   This latest record, however, is not the highest in the county's history, for in 1871 the people numbered 25,430.   This is doubtless owing to the fact that Radnor, like all the other agricultural counties of Wales, has suffered to some extent from the steady migration of the young people to the industrial districts.

But a census means much more than a mere counting of heads.   It is a decennial picture of the social life of the

Welsh Peasant Woman in National Costume

country.   Thus in 1901 it was shown that the 23,281 inhabitants of Radnor lived in 5068 tenements, of which 1302 contained less than five rooms each, and 3766 five or more rooms.   There were 4195 families residing in the county, and they averaged 4·8 members each.   No less than 77·6 per cent. of the population lived in rural, and 22·4 per cent. in urban districts, covering the county to the number of only 48 to the square mile, whereas Glamorgan was seen to have as many as 1065, and Wales as a whole 253 per square mile.   In 1911 the population of Glamorgan had risen to 1383 per square mile.

In the census of 1911 the males (and here Radnor-shire differs from many other counties) exceeded the females by 89.

In 1901 5146 children were of school age, i.e. between 3 and 14 years of age ; and there were 1609 persons over 65 years old.

Very interesting is the fact that (in the 1901 census, the latest available giving these details) 15,771 or 67·7 per cent. of the people of Radnor were natives of the county, that 2828 were born in the other Welsh counties, including Monmouth, and that London claimed 176, Ireland 80, Scotland 138, and the colonies 31 persons. Radnor has no barrracks, prison, or lunatic asylum.

The census showed that the Welsh language has become almost extinct.   In the Rhayader rural area there were but 47 persons who still habitually use Welsh alone, and in the rest of the county only four such could be found.

## 14. Agriculture—Main Cultivations, Woodlands, Stock.

When we reflect that Radnor is an upland county it is a matter of some surprise that it takes such a prominent place in Welsh farming. This can be accounted for in several ways.

(1) By the care bestowed on its best land.

(2) By the encouragement given to excel in farm work.

(3) By its freedom as a whole from such mines and manufactures as would be injurious to good agriculture.

History has given us some interesting pages on the state of agriculture in this county at various times; and perhaps we can better understand its present condition after a careful perusal of the records of the past.

Towards the end of the twelfth century Giraldus said of this neighbourhood (it was no shire in his time), "This country sufficiently abounds with grain; and if there be any deficiency it is amply supplied from the neighbouring parts of England. It is well stored with pastures, woods, and wild and domestic animals."

The gift tendered by Moll Walbee of Hay to King John's wife during the reign of that monarch was "400 kine all milkwhite except their ears which were tipped with red." This kind is claimed to be the parent stock of the modern Hereford breed of cattle.

Leland in the sixteenth century was enthusiastic concerning the productions of Radnor, more so, in fact, than of the neighbouring county of Brecon. He says : "Radenor wolle ys moch praised." "The valey about Radenor is veri plentiful of corne and gresse." "Presteine is a very good market of corne."

But when we come to the end of the eighteenth century we find John Clarke taking a very unfavourable view of the county's agriculture in his day. He does not hesitate to say that "the natural poverty of the soil is rendered still more unproductive from the uncommon indolence of the inhabitants, who cultivate the land in the same manner nearly as their ancestors did a thousand years ago." But he makes an important reservation in favour of the Hundred of Painscastle, for of this neighbourhood he says: "The superior industry and good management of the inhabitants as farmers have been successfully exerted to avail themselves of every natural advantage."..."They contribute in a considerable degree to rescue this much neglected county from a censure almost general."

About fifteen years after Clarke wrote was founded the Radnor Agricultural Society, which has done much to improve the husbandry of the county.

From the annual Government Returns for 1910 we learn that in that year 161,813 acres, or somewhat more than half the county, were under crops and grass. Of these, 39,100 acres, or less than one-eighth of the county, were arable, and 122,713 acres, or almost two-fifths, under hay of all kinds. The total corn area was 16,890 acres, of which oats alone accounted for 11,368 acres. Rye-growing evidently is not quite abandoned, for it was shown to be cultivated to the extent of 10 acres. Turnips and swedes occupied 5210 acres, potatoes only 690, and mangolds only 202 acres.

It is rather remarkable that although Hereford—the second hop county of England—adjoins Radnor, the latter grows no hops whatever. This has not always been the

A Radnorshire Ploughing Match

case, for Malkin, towards the close of the eighteenth century, saw near Cregrina, "a very fruitful hop-garden in the narrow flat between the river and the hill," and Williams, the county historian, spoke in 1810 of "the numerous and enlivening plantations of hops and cider fruit trees which decorate the vales of the Wye, the Somergil, and the Lug."

In the cultivation of the "cider fruit trees," for which Hereford is equally famous, Radnor, however, shows more emulation, for it grows the fruit to the amount of 690 acres, which is the third greatest apple acreage in Wales, coming next to those of Monmouth and Brecon. But little small fruit is grown; yet the terrier of Llanelwedd Church includes hops, flax, hemp, and honey as having

Sheep-Shearing in Radnorshire

once been objects of culture within that parish. It is certain that flax was once a common product of the county, for in 1270, Owen ap Goronwy, Constable of Builth Castle, sent the following letter to William Fitz-Adam, one of Mortimer's henchmen: "Know ye that if you do not permit Madoc to mow his crops *and cleanse his flax*, and harvest his hay as has been fully agreed upon, you choose war, and this beyond doubt ye shall have."

Radnor for its size takes a prominent place in Welsh wool and mutton, for in 1910 it was credited with 296,643 sheep, mostly of the Radnor, Cheviot, and Kerry stock. It also had 33,482 cattle, mostly Herefords, but butter and cheese-making are declining as regards quantity produced, for much of the milk is now sent daily by train to supply the populous centres of the Midlands and even of distant Lancashire.

Finally the farm produce of the shire is connected with 2223 holdings of various sizes over an acre in extent, 286 of which are owned or mainly owned, and 1937 rented or mainly rented. In the proportion of freeholders within the county, Radnor with a percentage of 12·87 takes third place in Wales; following Cardigan with 18·11 °/$_{o}$ and Anglesey with 13·71 °/$_{o}$. It is also interesting to note that it was amongst the first counties in the country to take advantage of the Small Holdings Act of 1907.

Much of the " stock " business of Radnor is carried on at fairs, of which every town and considerable village has a number held periodically. Knighton, dealing mostly in sheep and store cattle, has 18 fairs and six special sales for 1911. To transact this business it has a large sheep-market nearly two acres in extent. Rhayader also has 16 fairs for this year, the month of May alone accounting for four.

The chief horse-fair of the county is that of New-bridge-on-Wye, which is annually held about the middle

Horse Fair, Newbridge-on-Wye

of October. Llanelwedd, on the Radnor side of Wye, shares with Builth the prosperity attendant on its great cattle and sheep fairs, where in one year alone (1909) 45,204 sheep, 4469 cattle, 603 cows and calves, and 1637 pigs changed hands.

Radnor, like Brecon, has a great area of mountain and heath land, 131,755 acres, or nearly half its area, being returned as belonging to this class. Over all these mountains sheep and ponies roam in large numbers. The latter had attracted the notice of Leland, who traversed the newly-formed county in 1540, for he says, "In Melennith is a good breed of horse on a mountain called Herdoel." The raids and counter-raids on both sides of the Marches in ancient days were doubtless productive of much booty, if the sheep, ponies, and cattle of those days were as numerous and valuable as those that Radnor can show to-day.

## 15. Industries and Manufactures.

There is no more interesting chapter in the history of our country than that which tells of the rise and fall of industries. No such thing as absolute stability exists in any trade. The fluctuations that affect a particular industry may be caused in various ways. New conditions of labour, the competition of other districts, and even the changes in the habits of the people—especially in the matters of food and drink, dress and amusement—all these and more besides must be reckoned with in the carrying on of successful trade.

Radnor has always been at a disadvantage from an industrial point of view as it has no coal or iron, those powerful factors of a country's wealth. Neither has it any coast, nor even a navigable stream to induce capital to plant factories within its limits.

Its chief natural productions have always been wool

A Rhayader Tanyard

and timber, and it is with these that the few manufactures the county has ever had are mainly connected. Latterly a few industries in connection with its stone products have arisen, and later still steam and electricity have added a few more.

Oak being plentiful everywhere and the hills having always been good cattle-rearing ground, tan yards would naturally flourish. This is an old industry and every town in Radnor at one time was engaged in tanning and dressing hides. Rhayader has retained this industry until to-day, tanning being still its chief commercial activity. Apart from the hides and bark utilised within the shire itself, great quantities of both products used to be exported in their raw state to other counties.

Until 1820 glove-making used to be carried on near the eastern border, where " the industrious poor of not less than twelve parishes of Hereford and Radnor were afforded employment thereby."

But Radnor's most ambitious efforts in the manufacturing world have been in connection with the treatment of wool—its chief product. A cloth manufactory was established at Presteign as early as the reign of Elizabeth, and of it an old record states that " it suffered irreparable injury through the pestilence." This factory was latterly converted into a flannel-mill. Maestroyle, three miles west of Presteign, possessed a similar factory at the beginning of the nineteenth century, while Rhayader factory about the same time boasted " carding machines and spinning jennies " as a feature of its equipment.

Llandrindod in 1794 also commenced wool-weaving, but like the other attempts within the county the venture did not prove a permanent success. It was found that conveying the wool in its raw state to Yorkshire and the West of England paid better than dealing with it at home. Not that all attempts were utterly given up. Dressing and dyeing wool was carried on at Knighton for a considerable period later, and the same town still has a small woollen factory carrying on the trade.  Rhayader possesses another, but the wool-sorting establishment that existed in Glasbury about 60 years ago has long been idle.

The present state of the woollen trade within the county is certainly one of decay, and is not even as encouraging as it was in Malkin's day, for though he said that "there are no large manufactories established," he

Stone Crusher, Llanelwedd

also added, "but the people make a sufficient quantity of coarse cloth, flannel, and stockings for their own use."

The burning of limestone for agricultural purposes is comparatively an old industry and has been continuously followed at Old Radnor for more than a century; but the breaking up by machinery of stone blocks for road metalling is a modern one.  It is done by means of great steam crushers which can be regulated so as to give a

6—2

macadam of any required size. This industry gives employment to a number of people at Rhayader, Llandrindod, and Llanelwedd. At the last-named place the small chips of the rocks are utilised for making artificial stone blocks which are in great request for building.

The L. and N. W. Railway has an important railway depôt at Builth Road, employing about 100 skilled workmen for the upkeep of the long railway section between Shrewsbury and Swansea. Llanelwedd has a similar but smaller depôt serving the Cambrian Railway.

It will be noticed from the above that although Radnorshire has but few manufactures they are of a varied character.

## 16. Mines and Minerals.

Radnor is the poorest county of South Wales from a mining point of view. Not only has it no coal or iron, but there is no other mineral of such value as to induce operations on a large scale.

Lead-mining has been attempted from time to time and the county records even show this industry to have temporarily flourished in ages past. It is asserted that the old lead mine of the parish of Llandrindod was worked by the Romans, but no proof has been adduced to justify the statement. It certainly was worked as late as 1797. When we come to Stuart times we are on firmer ground, and there is no reason to doubt that Sir Hugh Middleton won much lead in the Cwmdauddwr district about the year 1640. The exact spot is still called Gwaith-y-Mwynau (the mine-works).

In 1796 an attempt was made to work the same ore at Cwm Elan. But, like a similar attempt at Llandegla, the mine soon became unremunerative and was in consequence closed.

To-day all the minerals that Radnor produces are worked in open quarries.

According to the latest statistics of the Home Office issued in 1910, it possesses ten quarries of various kinds, employing in the aggregate 95 hands, and producing in a year 12,440 tons of limestone, 24,506 tons of sandstone, and 2400 tons of igneous rock.

Roofing-tiles used to be quarried on Clyro Hill and in the Rhayader district, but at present there is no demand for them, and the quarries are closed.

Limestone has been worked for about two centuries in the east of the county, chiefly near the towns of Old Radnor and Presteign. The Old Radnor variety is of a greyish colour and is very suitable as a manure for agricultural purposes. The county records show that 22 kilns were in full work during the year 1819, and that up to the present time the output has been fairly maintained. The white limestone of the Presteign district is mainly used for building purposes, its cementing properties being superior to those of the grey kind.

The Llanelwedd quarries have lately come into prominence on account of their having supplied almost all of the massive masonry of the Birmingham reservoirs. The Llanfawr quarries of Llandrindod and the Llwynon quarries of Rhayader also produce a hard stone which is in great demand for road-making.

In various parts of the county are rich deposits of peat called in Welsh *mawn*. Since the depletion of the forests of the common lands peat-beds have been the chief fuel supplies of the shire for some centuries. They have

Gelly-Cadwgan Quarry, Llanelwedd

before now been a source of fierce litigation, scarcity of fuel on the one hand and feudal privilege on the other asserting each its imperious claim.

The best turbaries now worked are at Cwmdauddwr, Disserth, and Rhosgoch (Painscastle).

The saddest remains of former mining adventure in Radnor are those which mark the sites of the hopeless attempts at coal-winning. In spite of the teachings of geology men have been found at various times ready to risk their capital in this vain quest. The proofs of their failure are still to be seen near Presteign, Rhiwgoch (Nantmel), Weythel, and a few other places.

## 17. History of the County.

There is no separate mention in Roman Britain of the land we now know as Radnor. It lay for the most

Caractacus' Camp near Knighton

part within the district of the Silures, and shares the glory of Caractacus in his stand against the Roman legions. Caer Caradoc, for ever associated with his name, is the camp near Knighton where his last battle was fought.

That the Romans thought this district to be important

is attested by the number of roads and camps which they constructed within its bounds. When they left in the year 410 it was fiercely contested for by its various invaders between the fifth and the eleventh centuries.

Gwrtheyrnion, the wild region in the west of Radnor, is believed to have been named after Gwrtheyrn, better known as Vortigern, the wretched prince who first invited the Saxons to aid him against the incursions of the Picts and Scots. Tradition points to this district as the place to which he fled when his power was usurped by his Saxon guests. Be this as it may, we are certain that the Mercians were the untiring foes of the Welsh border tribesmen at all times.

The tale of Offa's Dyke from 750 A.D. to 1066 A.D. is one of unceasing warfare, when each nation held out grimly to prevent the transgression of the other. Gruffydd ap Llewelyn, in the time of Edward the Confessor, burst into the lowlands of the Western Midlands, sacked Hereford, and carried fire and sword throughout the country side. But immediately after we find Harold Godwin establishing an English colony west of the Dyke at New Radnor, and it is probably owing to him that we find such purely Saxon names as Kington, Huntington, Walton, Kinnerton, Harpton, Preston, Whitton, Norton, and Knighton interspersed with others of a distinctly Welsh character. Domesday refers to his possessions thus: "The King holds Radnor, Earl Harold did hold it."

Radnor never was a district unit in the sense that Ceredigion or Glamorgan was. These were governed by their own line of princes and formed small states in themselves long before the modern counties bearing their names were in existence as such. The shire on the Dyke, however, was always the prey of one or other of its strong neighbours, of Powys to the north, Mercia to the east, Dyfed to the west, and Morganwg to the south.

Before the year 1536 the history of this borderland was the history of its chief component lordships, namely, Gwrtheyrnion, Elvael, and Melenydd, of which the last was much the most important.

The Welsh lords of Melenydd were seldom other than princes of high degree, the title being often coupled with that of the Dukedom of Cornwall. But this line came to an end with Cadwallon ap Maelgwn, and then we find that William Rufus conferred the lordship upon Ralph Mortimer the Norman. This transfer to the conquering race was never acknowledged by the Welsh until Edward I confirmed the grant of Melenydd to Roger Mortimer of Wigmore. This territory of Melenydd anciently comprised a hundred townships, and extended into the district now called Montgomeryshire.

Another title often connected with the Dukedom of Cornwall was that of Lord of Radnor. The Breos family at times held it, and after the Mortimers and Breos' had intermarried the Lords of Melenydd became Lords of Radnor as well.

With the death of the last Llewelyn the Norman grip of the Wye country was tightened sevenfold. Whereas other parts of Wales were immediately after divided into shires, securing thus the benefits of strong law and regular government, Radnor remained in the heart of the Marches

Llewelyn's Cave, near Aberedw

to lie under the martial government of the Mortimers and Breos' until 1536.

Roger Mortimer, the favourite of Queen Isabella of England, had his estates confiscated, and Cantref Melenydd became royal property. But after another Roger Mortimer had equipped a body of natives and led them at Crecy, his honours were restored to him. So high stood the fighting prestige of the district at the time that, soon after, we find Edward III raising 2000 men in the Cantrefs of Elvael and Buallt to fight against the Scots. This prestige was still further enhanced by the valour of another body under Bryan Harley of Stanage at Poictiers.

But military prowess was no new characteristic of these mountain people. When the Archbishop Baldwin of Canterbury and Gerald the Welshman made their famous circuit of Wales for the preaching of the Crusades, it was New Radnor they fixed upon as the best starting-place for the purpose.

Richard II, being childless, announced the Roger Mortimer of his day to be his heir. But that far-seeing nobleman modestly declined the honour and went to Ireland as Lord-Lieutenant, where he was killed in a skirmish with the natives.

During the Glyndwr crisis, Radnor was the cockpit of the fighting in South Wales, its castles being taken and retaken again and again. It was at New Radnor Castle that Glyndwr hanged 60 of the garrison, and it was at Pilleth also that the "irregular and wild Glendower," through his lieutenant, Rhys the Terrible, inflicted the greatest defeat of all upon the English arms.

Henry IV and his gallant son Prince Hal often traversed the district in those troublous times. It was here doubtless that the latter learnt the art of war, which he afterwards so well exercised on the field of Agincourt, when he was loyally followed by Sir John Cornewall of Stapleton and Downton, William ap Llwyd of Trellwydion, and Sir Roger Fychan of Clyro—all tried and true men of Radnor.

The Mound, Pilleth Battlefield

Edmund Mortimer, who died without issue in 1425, was the last Mortimer who was both Earl of March and Lord of Melenydd. His sister Ann espoused Richard Plantagenet, whose son, Richard Duke of York, received all these estates and titles. He was killed at Wakefield Green in 1460. It was his son—the victor of Mortimer's

Cross—who as Edward IV instituted the Court of the Marches—that stern tribunal which by its inhuman treatment of the people created so much lawlessness on the frontier for the next 70 years. One of its presidents, Bishop Lee, is said to have hanged several thousands of the ignorant people during his period of office.

It can therefore be surmised with what joy the news was received in 1536 "that it is enacted that the Lordships, Townships, Parishes, Cwmwds and Cantrefs, of New Radnor, Elistherman, Elveles, Boughrood, Glasbury, Glandistry, Michael's Church, Moelienydd, Blaiddfa, Knighton, Norton, Presthend, Cwm-y-audwr, Rhayeder, Werthrynion, and Stanage shall be reputed and taken as parts and members of the county or shire of ' Radnor ' " (27 Henry VIII, c. 26).

The new shire of Radnor was considered to be of about the same grade and importance as Cardigan, Carnarvon, and Merioneth, for when Charles I made a levy of men for his Scottish war, Radnor had to contribute a quota of 50 men, which was the same number as that required of those counties. Its ship-money, however, was assessed at £490. 10s. 0d., which was somewhat less than the sum demanded of them.

During the Civil War the landed gentry as a rule sided with the King, but did not take any prominent action in his behalf, as Bradshaw of Presteign, the president of the regicides, overawed them with his energetic personality. New Radnor, nevertheless, was for a time held for the King.

The county towards the end of the seventeenth and

throughout the eighteenth century was the scene of the
labours of some of the leaders of Welsh Nonconformity.
Among them were Vavasor Powell, Walter Craddock,
and Howel Harris. The Wesleys, John and Charles,
were also frequent visitors. John Wesley's diary shows
that he ministered at Rhayader in 1750 and 1756.
Charles, his brother, married Sarah Gwynn, of the great
Garth and Llanelwedd family of that name.

"The Pales," Llandegley

The Society of Friends, consequent upon the visit of
George Fox in 1657, met with more success in Radnor
than in any other part of Wales. The quaint old meeting-
house, called "The Pales," is still standing at Llandegley,
a relic of the time when the Quakers were a power in
the religious and political life of the shire.

The last occasion on which Radnor became prominent before the outside world was during the Rebecca Riots of 1843–5, when the turbulence of the rioters and their frequent depredations made an unfavourable impression as to the conditions of life then obtaining within the county.

## 18. Antiquities—(a) Prehistoric and Roman.

No people can occupy a country or district for any length of time without unwittingly leaving marks of their occupation. In the absence of written records these remains are the only means by which the history and customs of early man can be ascertained. In various places ancient weapons, ornaments, utensils, and even carvings on bone, have been found, all of which tell us somewhat of the life of those far-off days.

Although South Wales as a whole, compared with Southern and Eastern England, cannot be considered as anything but poor in this respect, Radnor as a particular county has produced a considerable number of such relics, and the twentieth century has already added several items to its good record.

The prehistoric stone hammers or hatchets found in Wales have been very few in times past, that of Llanmadoc near Swansea perhaps being the most notable find. But November, 1910, saw the discovery of another similar weapon on Esgyrnderw Hill near Rhayader, where a good specimen was unearthed. It measured 6½ ins. long, and was perforated by a hole 1 inch in diameter. Another recent discovery was that of an object supposed to have been a prehistoric ornament or charm. It was found near Llandrindod in 1909. It is crescent-shaped, and measures 6 ins. from one horn to the other.

Of the larger remains of primeval man Radnor has many examples. Its barrows or mounds, round huts, menhirs (*meini-hirion*), cromlechs (*cromlechau*), stone circles, and camps are so numerous as to be quite a striking feature of the county.

Prehistoric Stone-Hammer
(*Unearthed near Rhayader* 1910)

Mounds, barrows, or *tumuli* are generally graves. Of these there are a great number in the county, no less than 30 being in the neighbourhood of Llanelwedd alone. A remarkably large one is Bedd-y-Gre in the parish of Llanddewi-Ystrad-Ennau. It has a foss all round it with a rampart circling that again. Another moated mound is Yr Hên Gastell in Aberedw. There is indeed scarcely a district in the whole shire which does not possess one or more of these tumuli. Rhayader, St Harmon's, and Stanage in the north, Blaen-edw and Llandrindod in the centre, Carneddau in the west, Court Evan Gwynne (Clyro) in the south, and the vales of the Lug and the Arrow in the east all have well-known specimens of these prehistoric graves, the circular ones at Carneddau being specially noted.

Some of the mounds, here and elsewhere, have been opened. Those of Llandrindod were found to contain

The Four Stones, Vale of Radnor

human bones and pieces of half-burnt charcoal, while another opened near Llanelwedd in 1906 revealed a *cistfaen*, or stone chamber, placed in the centre of a stone circle.

Other stone relics are the menhirs (*meini-hirion*, or long stones), of which specimens are seen at Cwmdauddwr

D. R.

(Carreg Bica), Glascwm, Abernant-y-Beddau, and Llan-drindod, while only recently another has been discovered in a hedge at Newbridge-on-Wye. The "Carreg Bica" of Cwmdauddwr is 7 feet 2 inches long by 12 inches wide, and has a cross roughly outlined upon it.

The celebrated Four Stones of the Vale of Radnor, thought by some to be merely erratic boulders from Hanter or Stanner Hills, are clearly the work of man, and there is another quartet of similar stones at Nant-y-Saeson, St Harmon's.

One of the chief circles of Wales is to be found on the crown of a dome-like elevation at Rhosfaen in the heart of Radnor Forest. It consists of 37 stones set in a circle 237 feet in circumference. The height of the stones varies from 2 feet 1 inch to 5 feet, and their distance apart from 2 feet 8 inches to 21 feet.

On one of the summits of Carneddau near Llanelwedd is a great cross marked in stones on the surface of the ground and pointing N.S.E.W., and on Caermaerdy farm (Glascwm), set in the midst of huge coarse stones, are the remains of the only cromlech of the county.

Celtic crosses, some of them showing beautiful "lace-work" design, exist at New Radnor, and Llowes. The last stands 7 feet 6 inches above the ground, and is 3 feet wide at the base. So much superstitious lore has grown round this stone in connection with the story of Moll Walbee of Hay, that the stone itself is now called by her name.

But perhaps the most interesting of all the prehistoric remains are the "camps," generally built around the

summit of a hill or bluff, fronted by a dry-stone wall as
a breastwork, and having a spring somewhere within its
boundary. Of these Radnor has an unusual number,
varying much in size, but all generally roughly circular
or oval in form. Those situated at Cwm-cefn-y-gaer,
Cefnllys, Beguildy, Burfa Bank, and Newcastle are the
best known. That at Beguildy, locally known as

Roman Camp, Castell Collen

Crug-y-Byddair, is traditionally connected with the great
Caractacus. Near it is a meadow known to this day as
"The Bloody Field."

Of traces of the Roman occupation Radnor has a
large number. For instance Roman tiles marked Leg. II
have been found near Llanelwedd, Roman coins and
bricks at Castell Collen, two leaden seals (one each of

Popes Honorius III and IV) at Abbey-Cwm-Hir, and a Roman urn at Brynllwyd. The last decade has added to this list, for some black ware marked Attili was found in 1909 at Castell Collen, and a cinerary or burial urn at Llandrindod in 1910. But the greatest discovery of all was at Nantmel in 1899, when a gold ring set with an onyx, a gold necklet, and a gold armlet were found in the rocks—all remarkably fine examples which are now in the British Museum. There is also a centurial stone in the church porch of Llanbadarn Fawr measuring 14 inches by 4 inches which is supposed to have been brought thither from Castell Collen.

Of Roman camps Radnor has some conspicuous examples. The Gaer, in Llanddewi-Ystrad-Ennau, Castell Collen or Cwm (also formerly called The Gaer), and yet another Gaer in the Vale of Arrow being the largest and best known. It was the great Welsh writer Carnhuanawc who latterly drew attention to the importance of the Cwm Station when it was in danger of being almost lost sight of in the thick growth of underwood.

The traces of the Roman roads are very plentiful, but these need not occupy us here as they will be more fully dealt with in the chapter on the Communications of the County.

## 19. Antiquities—(*b*) Irish and Saxon.

The only Irish remains within the county, apart from some place-names, are the foundations of the curious circular huts known as "Cytiau'r Gwyddelod," found on Carneddau, near Llanelwedd. In the neighbouring county of Brecon some stones inscribed with ogams—the old Irish characters—have also been discovered, convincingly proving the truth of the old records and traditions concerning the presence of the Irish in Siluria.

The Saxons, as has been mentioned, failed to make any permanent footing west of Offa's Dyke. The success of Harold in east Radnor was so recent at the time of the Norman conquest that it could scarcely be called a settlement, and Saxon remains cannot therefore be expected to exist in any considerable numbers to-day. The historical stones which Harold himself boastfully erected at certain points to declare that "Here Harold conquered," have long been lost.

The similarity of the decoration of the Moll Walbee stone at Llowes to that of another stone found in Durham Cathedral has occasioned some to think that this monolith might be Saxon, but this is mere conjecture.

Offa's Dyke itself is by far the most important Saxon relic, not only of Radnor, but of all the western shires. Constructed about the year 760 by the great Mercian king, it has figured in the history of the borderland ever since. Asser, the contemporary of Alfred the Great, wonderingly refers to it and to its amazing length "de mari usque ad mare." It has also been an important factor in the laws of both countries, because from its well-guarding only could safety be assured.

There is a general but erroneous impression that the Dyke was a military barrier similar to Hadrian's Wall or the Great Wall of China. It really was only a marked boundary line which could not be crossed in ignorance of

Offa's Dyke, near Knighton

its existence. It was undoubtedly well patrolled, and the towns and villages near it were compelled to equip a horseman for the chase of armed trespassers. By the law of Harold such trespassers if caught on the Saxon side were to lose their right hands, thus depriving them for ever of the power to bear arms.

The Dyke enters Radnor on its north-eastern frontier from the county of Salop. Passing through the town of

Knighton it runs for two miles almost in a straight line
to the south, and can be plainly traced through the parishes
of Norton, Whitton, Discoed, and Old Radnor. It then
enters Hereford, passing to the east of the town of
Kington, thus making that town with Huntington and
the more famous Hergest district entirely Welsh in ancient
days. Its highest point above sea-level is on Garbeth
farm between Newton and Knighton. The latter town
is called in Welsh "Trêf-y-Clawdd"—the town on the
Dyke. The section of the Clawdd near the town's golf-
links to-day shows it to be about 8 feet deep and 20 feet
across at the top. Other points where it is still conspicuous
are Evenjobb Hill and Burfa Bank, before it descends into
the Somergil valley to cross the stream at "Ditch Held."

Offa's Dyke has within the last ten years received
much attention. Among other interesting things discovered in connection with it is the fact that wherever it
is carried on a hillside the slope is always *from* the Mercian
border *towards* the Welsh territory. This was doubtless
of intent so that the Mercian sentry could overlook the
valley below. As a study in local corruption of place-
names it may be mentioned that the natives between
Knighton and Presteign variously designate the great
ditch as "Heyve Deytch," "Have Deytch," "Hoff
Deytch," and even "Half Ditch."

On the Beacon Hill in the parishes of Llangunllo and
Beguildy is a remarkable ridge of earth artificially thrown
up. It is called the "Short Ditch," but it is no part of
Offa's Dyke. It was possibly thrown up for defence by
a raiding party of either nation overtaken by a superior

force.  A similar dyke is to be seen about a mile from
New Radnor which is continued from one end of the
narrow vale to the other.

## 20.  Architecture—(a) Ecclesiastical.

About one-fourth of Radnor, consisting of the eastern
parishes from Knighton to Michaelstone, lies within the
diocese of Hereford.  The remaining three-fourths, consisting of the deaneries of Elvel and Melenydd in the
archdeaconry of Brecon, belong to the diocese of St
Davids.

Most of the churches of the county are plain and
ordinary in structure and are built almost exclusively of
native stone.  They often show the "cove" or "cradle"
roof and are surmounted in almost every case by a square
battlemented tower.

Although the churches are plain and unpretentious in
their outward appearance, Radnor being poor in good
building-stone, the woodwork throughout is of a superior
character, so much so that Radnor churches are noted
in this respect.  They generally have large chancels but
very often no aisles and no architectural division between
chancel and nave.

So much for general features.  A preliminary word
on the various styles of English architecture is necessary
before we pass to the special points of the churches and
other important buildings of our county.

Pre-Norman or—as it is usually, though with no great
certainty termed—Saxon building in England, was the

work of early craftsmen with an imperfect knowledge of
stone construction, who commonly used rough rubble
walls, no buttresses, small semi-circular or triangular
arches, and square towers with what is termed "long-and-
short work " at the quoins or corners.    It survives almost
solely in portions of small churches.

The Norman conquest started a widespread building
of massive churches and castles in the continental style
called Romanesque, which in England has got the name
of "Norman."    They had walls of great thickness,
semi-circular vaults, round-headed doors and windows,
and massive square towers.

From 1150 to 1200 the building became lighter, the
arches pointed, and there was perfected the science of
vaulting, by which the weight is brought upon piers and
buttresses.    This method of building, the "Gothic,"
originated from the endeavour to cover the widest and
loftiest areas with the greatest economy of stone.    The
first English Gothic, called " Early English," from about
1180 to 1250, is characterised by slender piers (commonly
of marble), lofty pointed vaults, and long, narrow, lancet-
headed windows.    After 1250 the windows became
broader, divided up, and ornamented by patterns of
tracery, while in the vault the ribs were multiplied.
The greatest elegance of English Gothic was reached
from 1260 to 1290, at which date English sculpture was
at its highest, and art in painting, coloured glass making,
and general craftsmanship at its zenith.

After 1300 the structure of stone buildings began to
be overlaid with ornament, the window tracery and vault

ribs were of intricate patterns, the pinnacles and spires loaded with crocket and ornament. This later style is known as "Decorated," and came to an end with the Black Death, which stopped all building for a time.

With the changed conditions of life the type of building changed. With curious uniformity and quickness the style called "Perpendicular"—which is unknown abroad—developed after 1360 in all parts of England and lasted with scarcely any change up to 1520. As its name implies, it is characterised by the perpendicular arrangement of the tracery and panels on walls and in windows, and it is also distinguished by the flattened arches and the square arrangement of the mouldings over them, by the elaborate vault-traceries (especially fan-vaulting), and by the use of flat roofs and towers without spires.

The medieval styles in England ended with the dissolution of the monasteries (1530–1540), for the Reformation checked the building of churches. There succeeded the building of manor-houses, in which the style called "Tudor" arose—distinguished by flat-headed windows, level ceilings, and panelled rooms. The ornaments of classic style were introduced under the influences of Renaissance sculpture and distinguish the "Jacobean" style, so called after James I. About this time the professional architect arose. Hitherto, building had been entirely in the hands of the builder and the craftsman.

When the early builders wanted a larger place of worship they did not pull down all the work of their forefathers but added to or altered it. Hence we often find a portion of a church built after one style while another portion or portions belong to a wholly different period.

Radnor, like other counties, has churches, or portions of them, representing every period we have mentioned. There is no doubt that several pre-Norman churches once occupied the sites upon which the present-day churches of the county stand. This is notably the case where the buildings were first dedicated to Welsh saints, e.g. Glascwm (St David), St Harmon's (St Garmon), Llanbister (St Cynllo), Llanbadarn Fawr (St Padarn), etc.

The churches of Glascwm and St Harmon's are specifically mentioned by Giraldus Cambrensis in his well-known *Itinerary*, and are there spoken of as old sanctuaries, around which several observances had in the course of time become recognised as customs. Now as Giraldus wrote this about the year 1188, i.e. when the Norman or Romanesque Period was drawing to a close, it is almost certain that the original churches of these places were pre-Norman. There is an undoubted relic of the same period within half a mile of the present Llanelwedd church, where the foundations of the ancient Celtic church, dedicated to St Elwedd, are still to be seen.

Llanbadarn Fawr has a celebrated tympanum, which with its porch form the only Norman work, that can unmistakably be pointed out as such, now existing within the shire. The arch was probably built about 1150–1160, its feature of out-turned zigzag moulding proclaiming it to be late Norman work. Tympana are rare in Wales,

Norman Doorway with Tympanum, Llanbadarn Fawr

the only other instance being at Penmon church, Anglesey.  The Llanbadarn tympanum is a very fine specimen,
and shows to what perfection of skill the masons of the
twelfth century had attained.  Portions of St Cynllo
(Llanbister), particularly of the nave, are also thought
to belong to the Norman period, and there can be little
doubt that a Romanesque building preceded the fourteenth century church recently restored at this place.

It is not many of the Radnorshire churches that can
be ascribed to the Early English period, the great wave
of activity that characterised English church-building of
this time having evidently stopped short of the remote
highlands of Elvel and Melenydd.   Pilleth church, however, has a tower which is undoubtedly thirteenth century,
and the single lancets of both nave and chancel of Bleddfa
point to the style of work usually associated with that age.
The arcade of five pointed arches now incorporated in
the church of Llanidloes, within the neighbouring county
of Montgomery, forms an interesting relic of Radnorshire
architecture of the twelfth and thirteenth centuries, for
when Abbey-Cwm-Hir shared the fate of the other
monasteries (1535–40), these beautiful features of the
building were transferred to Llanidloes, where they still
remain.

Of the Decorated style of building, which came in
about 1300, the county has numerous examples, the east
windows of Llanbadarn Fynydd and Whitton, another
window at Pilleth, and portions of Presteign church, clearly
belonging to this style.   But perhaps the church com-
pletest in its Edwardian character is that of Clyro-on-Wye,

where the greater part of both chancel and nave displays the style characteristic of that half century.

When towards the close of the fourteenth century the country recovered from the ravages of the Black Death, which had wrought such havoc throughout our land, Radnorshire, in common with the rest of the country, showed great activity in the building of new, and the re-building of old churches. This was the beginning of the Perpendicular period, exemplified within our shire by the fine east windows of Gladestry and Disserth, and the St Catherine's wheel of Old Radnor. Glascwm church, mentioned above, was evidently rebuilt about this time, for it has several good Perpendicular windows dating from that revival.

Several Radnor churches have structural features of note. The oak screens of Old Radnor, Llanbister, Llanano Cefnllys, and Aberedw are beautifully carved. Llanbister's tower is placed at the east end of the church, and the chancel of Aberedw is lower than the nave.

The font of Old Radnor is probably one of the oldest ecclesiastical relics now used in connection with public worship within the kingdom. It has long been the puzzle of antiquaries, who however agree as to its pre-Norman character. The organ-case of the same church is believed to be the only survival of its kind in the country. It dates back to the Gothic period.

As compared with these vestiges of long past centuries, any structure connected with Nonconformity appears as of yesterday, but it may be stated that Llwynllwyd and Maes-yr-onen, both near Glasbury, are among the oldest

Rood Screen, Llanano Church

Nonconformist places in the Principality, and "The Pales," a thatched building in the parish of Llandegley (p. 94), has been a chapel of the Society of Friends (once numerous in Radnor) for about two hundred years.

Before the Reformation, religious houses—chiefly abbeys, priories, and friaries—were very common throughout the country. Not only were the monks the leaders

The Font, Old Radnor Church

in religious matters, but they led in learning, agriculture, medicine, and other branches of knowledge as well. Radnor possessed one abbey—that of Cwm Hir—dear to the heart of every Welshman on account of its being the resting-place (according to the Chronicles of Chester and Worcester) of the headless body of Llewelyn the Last. It was founded in 1143 by Cadwallon ap

Madog, the last of the Welsh lords of Melenydd, and was an offshoot of the Cistercian abbey of Whitland. It appears to have been built on the usual plan of the Order, and the original intention, if carried out, would have made it one of the largest abbeys in England and Wales. But instead of the 60 monks it was intended to maintain, it had but three at the time of the dissolution of the monasteries. In fact the abbey was never rich,

Maes-yr-onen Chapel, Glasbury

because the Welsh princes ceased to be lords of Melenydd about the time of its foundation, and the Mortimers who succeeded them lavished their favours on Wigmore Abbey in preference. In the time of Glyndwr the majority of the brethren were Englishmen, who during the struggle naturally sided with the king. For that reason the Welsh chieftain wrecked their house in one of his forays into the neighbourhood. But it does not appear to have been totally destroyed by him, for in the Civil War it underwent a siege in 1644, when Sir Thomas Middleton, the Parliamentary general, stormed it, "taking 70 prisoners and much store of ammunition."

Now, however, there is left no arch, window, door, or column on the site of this famous edifice. We must seek them elsewhere, where the pointed arches of Llanidloes church, the screens of Newton, Llanano, and Llanbister, the woodwork of Ty Faenor House, and many farms in the district have been embellished by the beautiful handiwork once adorning Abbey-Cwm-Hir.

Very interesting are some of the records of this abbey. The abbot owned a sheep-walk on Llechell Wiham Common, and possessed five granges besides—two in Radnor, "Clyro Grainge" and "Pepyll Wyllt," and three in Montgomery. Five mountains—Clôg Arthur, Withon, Berfath, Allt Mynachlog, and "another"—were also owned by him, showing upon what a large scale the farming operations of the abbeys were conducted. For while the Benedictine monks paid for labour, the Cistercians made a point of being self-supporting. Abbey-Cwm-Hir, therefore, doubtless had its mill, smithy, well, and bakehouse in the immediate vicinity, for the brethren not only grew their own corn, hay, flax, wine, and wool, but also made their cloth, fulled it, and tailored it.

Near Rhayader bridge the Blackfriars once had a cell, but very little is known of it or its temporalities beyond the fact that the cell was afterwards a dwelling-house, which had some subterranean passages leading from it.

Capel Madoc, another cell in the parish of Cwm-dauddwr, was an outpost of the Strata Florida Abbey in Cardiganshire.

In several parts of Radnor there are ancient farm-houses called Monachty, i.e. The Monks' House. These

Ruins of Abbey-Cwm-Hir

were probably granges which belonged to one or other of the religious houses of South Wales or the Border Counties, but which on occasion served for other purposes. The Monachty of Pilleth has a dungeon and a large hall, which seem to verify the local tradition that justice was occasionally administered there.

## 21.  Architecture—(*b*) **Military.**

We have already noticed that the British and the
Romans erected many strong fortifications within the
county, and that even the churches at times were built
with a view to defence.  In the next chapter we shall
have something to say about certain manor-houses that
answered the same purpose.  But the most perfect buildings from a military point of view were undoubtedly the
Norman castles, of which Radnor, in proportion to its
size, had a larger number than any other shire within the
four countries.

The best land in the county, as we have seen, lies
almost wholly within the valleys of the principal streams,
and this it was, no doubt, that led the Norman adventurers
to conquer Melenydd.  When it fell into their hands
they forthwith built strong castles to retain their hold.
Almost without exception therefore we shall find the
castles dominating the most fertile portions of the vales.

These strongholds had much in common in the plan
of their construction.  The usual type had a towering
keep or donjon, with inner and outer baileys or yards,
surrounded by a high wall.  The massive gateway was
entered by the drawbridge over the moat which, when-
ever water was obtainable, encircled the castle.  There
were also outbuildings within the yards, a chapel and
courthouse, as well as the necessary stables and smithy.

Of the numerous castles the county once possessed
there is not one to-day which has any of its parts intact,

**Sketch Map showing the Chief Castles of Wales and the Border Counties**

the stern warfare of earlier times and the vandalism of
modern days having all but obliterated their walls.   The
mound, the foundations, and portions of the moat all
littered with debris, are usually the only marks left to tell
their story to the coming ages.

The oldest as well as the most important of these
great forts was undoubtedly the castle of New Radnor,
originally built by Harold Godwin on Radnor Tump in
1064 to guard the colony recently planted by him in the
vale below.   As this castle was erected before the Norman
conquest, and, moreover, by the hand of a Saxon of
Saxons, it has been denied by some to have been a fortress
of the Norman type at all.   But when we remember that
Harold had visited Normandy before the date of its
erection and had as a warrior doubtless noted the great
military structures there, and especially as we are certain
that Ralph Mortimer in 1091 had greatly enlarged and
strengthened its defences, New Radnor must undoubtedly
be described as the first Norman keep in Wales.   It was
probably also the last in the county to keep its walls, for
the final spoliation of the material of its sturdy edifice was
only made within the memory of living man.

This old fortress played a stirring part in the wild life
of the marches, and it was quite in keeping with the
warlike reputation of the neighbourhood that Archbishop
Baldwin and Archdeacon Giraldus should first preach the
Crusades here.   Glyndwr found it a difficult place to
conquer, and in order to strike terror into the hearts of
his enemies he hanged sixty of the garrison on the castle
walls.

The castle underwent another siege during the Civil War, but gunpowder and cannon having then come into general use the garrison was soon compelled to capitulate to the victorious Parliamentarians. That the castle was of large extent is shown by Speed's map, one of the oldest Welsh maps in existence, which also shows that a large area was occupied by the town of New Radnor, which was walled in the Middle Ages. Its defences were at least partially in existence as late as 1540, for Leland in that year said, " In the walle appere the ruins of four gates."

Another interesting Radnor fortress was Painscastle, so named after its founder, who built it in 1130 in the parish which is now named after the castle itself. He was a comrade of the first Ralph Mortimer, a strenuous fighter, and one of the Crusaders who fell in the Holy Land. This is the Welsh castle so often mentioned in Scott's *The Betrothed*, under the name of *Garde Doloreuse*, and the particular siege described therein is the one it underwent at the hands of Prince Gwenwynyn of Powys.

Colwyn Castle was built a few miles above Aberedw by William de Breos in 1216, and by him was given over to his wife, the strong-minded Moll Walbee of Hay. For this reason it is still sometimes called Maud's Castle. Under the shadow of its walls Llewelyn the Last fought and defeated Sir Edmund Mortimer, who afterwards died of his wounds.

Cefnllys Castle, almost surrounded by the river Ithon, was built in 1242 by the Ralph Mortimer of that day. Its chief claim to notice is the fact that the Court of the

Marches was often held there in the fifteenth and sixteenth centuries.

Another castle on the Ithon is that of Dinbod, built by either Talbot or Tibbelot, lieutenants of the first Ralph Mortimer. It vies with Carreg-cennen of Carmarthenshire or Harlech of Merionethshire in wildness of situation, for it is built on Crogen Hill, 1250 feet above

Ruins of Dinbod Castle

sea-level, immediately overlooking the steep banks of the river. It was practically destroyed by Llewelyn ap Gruffydd in 1260, but up to the beginning of the nineteenth century a portion of its keep was still standing.

Boughrood Castle in the Wye Valley was in ruins long before Glyndwr's time. It was one of the first mortared castles built by a Welsh prince, its construction being credited to Einon Clyd, the Lord of Elvael, murdered by the Normans at Cwmdauddwr, whose assassination is commemorated by the crossed stone still standing in that parish.

In order to guard the Wye Valley from depredations

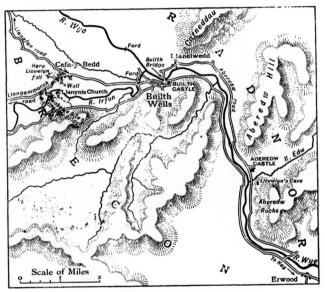

The Scene of Llewelyn's Downfall

from the north Rhayader Castle was built in 1178 by another Welsh prince, Rhys ap Gruffydd, better known as the Lord Rhys.

Up and down the county there are sites of various other fortresses about which little or nothing is known.

Such are Knighton, Knucklas, Ffaled, Matchway, Cymaron, and Aberedw Castles. Of the last-named, for instance, nothing beyond the fact of its having been a hunting-seat of the last Llewelyn can with certainty be stated. It is conjectured by some that when on his hunting expeditions he had at various times visited his old comrade of the Battle of Lewes—John Giffard—at the neighbouring castle of Builth, just beyond Wye. This was used to lure him to his destruction, for when he left Aberedw for the last time he travelled towards Builth, expecting to be received with open arms as usual. For that reason he had taken with him but a small retinue, and thus became the victim of a foul plot in which Giffard, Roger L'Estrange, and Roger and Edmund Mortimer were the chief actors. Entrance to the castle being denied him, he and his trusty followers had no chance in the open country and all fell victims to successful villainy. Giffard and L'Estrange received large estates in Gloucester and Norfolk from Edward I for their share in the massacre.

It is worthy of note that the large tract of land west of Rhayader never possessed a castle of any kind. The poverty of the soil probably accounts for this.

## 22.  Architecture—(c) Domestic.

During the period of the castles and for many ages prior to it the peasantry of this country had to be content with habitations of wood and even wattle or plastered osier work. These were naturally very fragile, and it is no wonder that few, if any, survive. But the presence of such fine stone structures as the castles, abbeys, and churches had its effect in time on the lowlier buildings. It took some ages, however, before stone-built houses became common. It was the landed gentry who led the way, being doubtless encouraged by the less frequent

**"The Stones" Farmhouse**
(*Where Charles I was entertained*)

occurrence of war after the Battle of Bosworth (1485). They wanted to be free from the discomforts of a castle, and at the same time to be ready in defence should an emergency arise. The result was the Tudor Manor House, a building which was sometimes moated and always capable of sustaining a siege of some duration.

Of this class of habitation—half fortress, half grange—Radnor has a large number. They are usually large in size, thick-walled, many-gabled, and low in roof, but almost always stately in appearance. There is hardly a district in the whole county that has not a few of these interesting buildings. Their oaken beams and solid masonry give them a strength that the builders of the twentieth century do not rival. Spacious fireplaces, over which rise mantels of elaborate workmanship, are the rule. The windows are generally low and small-paned, but the doors combine strength and ornament to a high degree. Very few of these houses are slated, the majority being roofed with that native stone which Welsh poets of the period are so fond of describing as " to teils."

Such were the houses of the great from the fifteenth to the eighteenth centuries when the owners of the soil usually farmed their own land.

The following incomplete list shows how rich Radnor is in buildings of this character:

| LLANELWEDD DISTRICT | | | | | |
|---|---|---|---|---|---|
| | Llanelwedd Hall, | the home of the | Gwynne family | | |
| | Wellfield | ,, | ,, | Thomas | ,, |
| | Pencerrig | ,, | ,, | Jones | ,, |
| | Pantyblodau | ,, | ,, | Crummer | ,, |

| RHAYADER DISTRICT | | | | | |
|---|---|---|---|---|---|
| | Cwm Elan | ,, | ,, | Shelley | ,, |
| | Nantgwyllt | ,, | Shelley and Lewis Lloyd family | | |
| | Doldowlod | ., | ,, | Watt | ,, |
| | Noyadd | ,, | ,, | Powell Evans | ,, |
| | Rhydoldog | ,, | ,, | Oliver | ,, |
| | Dderw | ,, | ,, | Prickard | ,, |
| | Llwynbarried (Nantmel) | | | | |

| | | | |
|---|---|---|---|
| **RADNOR DISTRICT** | Harpton | the home of the | Lewis family |
| | Downton | ,, ,, | Percival Lewis ,, |
| | Hergest Court | ,, ,, | Gwyn Vaughans ,, |
| | Gladestry | ,, ,, | Meyrick ,, |
| | The Stones | | |
| **GLASBURY DISTRICT** | Solers Castle | ,, ,, | Solers ,, |
| | Trewelley (Llowes) | ,, ,, | |
| | Old Clyro Court | ,, ,, | Vaughan ,, |
| | Gwernfythen | ,, ,, | Whitney ,, |
| | The Screen | ,, ,, | Vaughan Williams ,, |
| **PRESTEIGN DISTRICT** | Boultibrook | ,, ,, | Jones-Bridges ,, |
| | Evenjobb | ,, ,, | Harleys ,, |
| | Trewern | ,, ,, | Hartstongue ,, |
| **KNIGHTON DISTRICT** | Monachty (Pilleth) | ,, ,, | Price ,, |
| | Pilleth Hall | ,, ,, | Price and Lewis ,, |
| | Farrington | ,, ,, | Culter ,, |
| | The Great House | ,, ,, | Crowther ,, |
| | Stanage House | ,, ,, | Rogers ,, |
| | Llanddewi Hall | | |
| **LLANBISTER DISTRICT** | Llynwent | ,, | ,, Vaughan and Meredith ,, |
| | Croes Cynon | | |
| | Ty Maenor | ,, ,, | Fowler ,, |
| | Crychell | ,, ,, | Evan Stephens ,, |

A few of them are still inhabited by descendants of their former owners, but many more have been turned into farmhouses.

The story of these manor-houses carefully and worthily written would make a most interesting volume.

Charles I was entertained at The Stones. Hergest Court (east of the shire's modern boundary but west of

Offa's Dyke) has given Wales one of its oldest and most-prized books *Llyfr Coch Hergest*, and Byron's *Ianthe* lived at Kinsham Court in the same district. It was to Doldowlod that Watt retired after making the whole world indebted to him, and it was at Cwm Elan and Nantgwyllt that Shelley found inspiration for some of his finest work.

Maesllwch Castle, Glasbury

Many of the mansions that have arisen since that time fully sustain the great reputation of the past, Maesllwch Castle, Abbey Hall, Newcastle, and Boughrood Castle being representative of a large number of perhaps equally beautiful country seats.

The farmhouses and cottages of the county are also, generally speaking, of a spacious and therefore healthy character. Malkin, over a century ago, said, " Their

cottages in general seem to be substantially weatherproof,"
and the favourable remarks by him and by more recent
writers have been strikingly borne out in the reports of
the war against consumption now being fought. Tuber-
culosis is known to be largely a question of housing, and
when the accounts of its ravages in the different counties
had been tabulated Radnor was found to suffer less in

Nannerth Farm House, near Rhayader

proportion than any other county in South Wales.
Cardigan, which is the shire that most resembles Radnor
in everything but housing, was found to have the highest
death-rate from consumption.

Many of the Radnor farmhouses still remain very
much as they were built in Elizabethan times. One
authority says, " They were rude but substantial dwellings
constructed of the large schistose flagstones of the district

with a chimney shaft of some pretensions terminated by string courses." Interesting examples of such houses are found in the Rhayader district, where Nannerth Ucha, Nannerth Ganol, and Cwm-Coel farms in particular claim attention. Portway (Bryngwyn) is known to have been farmed by the same family for four centuries, and Bryn-llwyd (Glascwm) is claimed to have belonged to the

The " Radnorshire Arms," Presteign

Prices for a thousand years. The Old Clyro Court farm still has an embattled gateway and arch as an approach to the house, and the Court Farm (Edw Valley) is pointed out as the ancient home of the great Baskerville family, while the traditions of the Lower Heath farm, near Presteign, show that Charles I made it his home for a time after Naseby had clouded his fortunes.

It is worthy of note that oaken walls still do service in some of the Radnorshire public edifices, an inn in the main street of Rhayader and a portion of the Town Hall of the same place giving evidence of the fact. Several of the inns of the county are of great antiquity and of no little interest. The well-known "Forest Inn" (Llanfihangel-Nant-Melan) and "Beggar's Bush" (Discoed) are both ancient; "Rhydspence" with its wonderful oak carving is one of the sights of the Whitney district; and the "Radnorshire Arms" at Presteign is known to have belonged to Bradshaw the regicide.

## 23. Communications—Past and Present. Roads, Railways.

We know nothing of the roads of our province before the subjugation of Caradoc in the year 51 A.D. by the Romans, who hastened to connect their chief stations. The nearest to Radnor of these great arteries were the Julia Montana from Caerlleon-on-Usk to Caerbannium (Brecon) and the Sarn Helen from Neath to Chester. The principal forts of the Radnor district were connected with these main roads, and with each other by means of cross roads of similar structure.

Caerfagu or Castell Collen, Llechryd, and the Gaer (Michaelchurch) were among the chief Radnor stations of the Roman era, and a causeway connecting the three undoubtedly passed from Caerfagu to Llechryd via Disserth, and thence through Llanelwedd and the uplands of

Llansantffraid, Glascwm, and Newchurch to join the Gaer on the Arrow in East Radnor.

A smaller road joined Caerfagu to Llanelleonfel in Breconshire over the Wye ford of Newbridge, and another connected it with Caersws in the north.

Michaelchurch-on-Arrow was also connected with the Julia Montana to the south by means of a branch via Clyro and Abergavenny, which crossed the Wye near the present Hay bridge.

All these roads were excellently made and served the country for many centuries. But after the Romans left, road-making seems to have become a lost art, for during the long period between 410 A.D. and about 1790 the dwellers of the land—native and alien alike—not only did not construct new roads, but neglected to keep the excellent Roman causeways in ordinary repair. Williams, the historian of the county, said that about the year 1760 "the roads were mere gullies worn by torrents," while John Clarke declared in 1794 that "the parochial roads were bad and those called turnpike much worse."

But the close of the eighteenth century saw the commencement of a great change in this respect. In the years 1791 or 1792 the Radnor Turnpike Road Act was passed, to connect the county with Cheltenham through Hereford, and with Aberystwyth through Rhayader. This road was well made and the coach travelled it for many years.

That the improvement of the roads, although steadily persisted in, was a slow process is proved from the Rebecca Riots of the Forties, when the Radnor roads came in for much attention. One witness before the Government Commission of 1844 said "it was a grievance to have to pay for a road which was a perfect basin, and held water and mud up to the horse's belly." The Rhayader district was much disturbed about this time : several gates had been destroyed, and in the town of Rhayader itself the presence both of policemen and soldiers was found

Knucklas Viaduct, L. & N. W. Railway

necessary to support the collection of the tolls at the gates, which were exceedingly unpopular as they were higher here than under any other trust in South Wales.

Radnor to-day has roads that can compare with any in the Principality, and this fact has been found in this day of motor-cars to have been a splendid investment for the prosperity of the county.

Radnor has no canals, but its railways, in proportion to its size, are numerous.

The London and North-Western Railway crosses the shire from Builth Road to Knighton, serving on its way the important district of Llandrindod Wells. The whole of the southern boundary is skirted by the Midland Railway from Hereford to Brecon, and the western side by the Cambrian system, which has stations at Boughrood, Aberedw, Llanelwedd, Builth Road, and Rhayader.

The eastern boundary is also pierced by two branches of the Great Western Railway, one serving Presteign and the other New Radnor.

A mineral railroad was made about the year 1818 to connect the Burlinjobb limestone quarries with Brecon Canal by way of Hay. But the period of its utility has long passed, as it could not compete with the superior locomotion of the modern railway.

## 24. Administration and Divisions— Ancient and Modern.

Under the rule of its native chiefs the Radnor lordships, like other Welsh lordships, were divided into hundreds, each of which was again subdivided into two or more commotes or *cymydau*.

The government of the lordship was carried out by means of different courts wherein justice was meted out in accordance with the Laws of Hywel Dda, codified about the middle of the tenth century. The ordinary

law court, which also constituted the principal instrument
of local government, was the Court of the Commote.
But mention is also made of the Prince's Court, Court
of the Cantref, Special Court, and Ecclesiastical Court.

The principal officers engaged were the *brawdwr*
(judge), *canghellor* (chancellor), *maer* (mayor), *ysgolhaig*
(secretary) and *rhingyll* (sergeant)—all nominated by the
ruling prince. Professional lawyers were not unknown,
for we find the names of *cynghaws* (pleader) and *canllaw*
(guide) in connection with particular cases.

The courts were usually held in the open air and
the pleadings were mainly oral. When the formalities
of the constitution and opening of the court had been
minutely observed, a priest prayed, the judges recited
their Paternoster and the witnesses were sworn on relics,
this mode of swearing being of all others considered the
most binding on one's conscience.

The weight of the evidence tendered depended to a
great extent on the social position of the witness. The
word of an *alltud* (foreigner) was worthless against a
Welshman of pure descent ; no wife could give evidence
against her husband, and apostates, perjurers, and common
thieves had for ever forfeited the right to bear witness at
any time.

The punishment was borne in fines, there being a
complete system of assessment for all possible offences.

At a later period we find that the commote court,
instead of being held in the open air as formerly, was
held within the castle, the dungeon became the prison,
and the Norman retainers officers of the Court. These

underlings had no sympathy with the Welsh and often oppressed them unmercifully.

Relics of these days are the Courts of the Manor still held in various parts of the county of Radnor. Although in the twentieth century they have almost become mere names, in Norman times they played no inconsiderable part in the adjustment of rights and privileges as between lord and tenant in connection with the land and the forest.

There is nothing more confusing than the constitution of the various parishes of our county. Subdivision and changes of boundary are constantly going on, the "parcel" of one decade may be the independent parish of the next, and a number of parishes may often be amalgamated to make one larger authority.

About the year 880, when Rhodri Mawr was Prince of all Wales, the land now called Radnor was divided into three cantrefs, viz.

| Melenydd | subdivided into 4 commotes, |
| Elvel | " " 3 commotes, |
| Y Clawdd (i.e. The Dyke) | " " 3 commotes. |

When Henry VIII formed the new county, it contained 52 parishes arranged in six hundreds. The number of the parishes remained constant until 1809, but by the 1911 Census they had increased to 63.

The county is also divided into parishes of another kind—the ecclesiastical—of which it has 46.

The chief officers of the county are the Lord-Lieutenant and the High Sheriff. The former is appointed by the sovereign for life, and is generally a nobleman or great landowner. The High Sheriff used to be appointed for life also, but now his term of office lasts a year only, the new High Sheriff being chosen every year on the morrow of St Martin's Day.

Great changes were made in county government in 1888, when the various bodies governing the police, high-

The Town Hall, Rhayader

ways, asylums, etc. were merged into one County Council, consisting, in Radnor, of 24 councillors and eight aldermen, the latter being chosen from the elected councillors or from suitable outside men without an election. Since that date the management of county education, elementary and secondary, has also been given to them.

In 1894 there was a further re-arrangement in local

government, when the Parish Councils Act came into force. The more important parishes were then formed into Urban District Councils, and the remainder were grouped into a number of Rural District Councils. Each rural parish again had a little parliament of its own given to it, wherein its more immediate needs were discussed and carried out.

Knighton

By this Act, Radnor was granted three Urban District Councils—Presteign, Knighton, and Llandrindod Wells— and five Rural District Councils—Colwyn, Knighton, New Radnor, Painscastle, and Rhayader.

Roughly coinciding with the Rural District Council areas are those of the Poor Law Unions, two only of which—Rhayader and Knighton—are centred within the county. The remaining portions of the shire are allocated to Unions centred in other and neighbouring counties, the overlapping being comprised of the several districts covered by the Rural District Councils of Colwyn, New Radnor, and Painscastle above. The total rateable value of these divisions for the incidence of the county rate was, in 1909, £260,034.

Radnor has one Court of Quarter Sessions, and is divided into seven Petty Sessional Divisions to try and to punish petty offences against the law. This duty is carried out by local magistrates, of whom the county has about 105 at the present time.

Radnor has now no corporate town, but New Radnor once possessed a charter given to it by Queen Elizabeth.

Before 1886, this town (with the contributory boroughs of Cefnllys, Knighton, Knucklas, Presteign, and Rhayader) had separate representation in Parliament, but at that date they were all merged in the county to elect one member for the whole area.


## 25.  The Roll of Honour.

" *Ymhob gwlad y megir glew* " (" Every country breeds great men ") is a popular as well as a true proverb, and Radnor, sparsely populated as it always has been and now is, has many names of which it is proud. In this chapter, however, we can select only a few of the more prominent and representative.

Although the history of the county deals largely with

kings and princes, viz. Henry III, Henry IV, Henry V,
Edward IV, Charles I, Llewelyn the Great, Llewelyn
the Last, and Owen Glyndwr, it was only from stress
of war or politics that they entered the county.

The third Henry kept court at Painscastle in 1231,
and Charles I was glad to find a refuge in the wilds of
the Border after the fateful battle of Naseby. A less
trustworthy story says that his son Charles II followed
his example after the battle of Worcester, and traversed
the eastern part of the county in the company of the
faithful Penderel.

Radnor was often visited by Edward IV in his younger
days, when as Earl of March he was heir to the great
house of Wigmore.

This shire also saw the last and saddest chapter in the
life of Llewelyn ap Gruffydd as it did of the brightest in
Glyndwr's meteoric career.

The Lord Rhys of Dyfed had as much to do with the
making of the history of Rhayader and Painscastle, as his
kinsman and contemporary, Giraldus Cambrensis, had in
the writing of it. The former led a strenuous life in
endeavouring to stem the Norman invasion, and the latter
no less strenuously fought for the independence of the
Ancient British Church.

Radnor has given several divines of note both to the
Anglican Church and the Nonconformist bodies. Thomas
Huet, Rector of Cefnllys and Disserth, helped William
Salisbury in 1567 to translate the Greek Testament
into Welsh, and W. Jenkin Rees—a clergyman who
lived nearly half a century in Radnorshire—produced

the *Lives of the Cambro-British Saints*, a work of great merit.

Vavasor Powell, one of the pioneers of Welsh Nonconformity, was born at Knucklas, and William Williams

Cwm Elan House, Shelley's Residence
(*Now submerged*)

Pantycelyn, its sweetest hymn writer, received his education at Llwynllwyd, near Glasbury. Of the modern school of Welsh Nonconformist pulpit orators Radnor has been connected with two of the most famous, for

Thomas Jones of Swansea, whom Browning so much admired, was born at Rhayader ; and J. R. Kilsby Jones, who at one time was one of the most noted men in Wales, ministered to a Llandrindod Church. A very telling pen-picture of him is to be found in Miss Braddon's book, *Hostages to Fortune*, which deals largely with Llandrindod life of 40 years ago.

Mention has already been made of Shelley's connection with Cwm Elan, and of Byron's visits to Presteign. But this does not exhaust the list of Radnor's literary connections by any means, for its native poets and prose-writers, although not commanding such a large circle of admirers as those great names, still have done good service in a modest way. Dafydd y Coed (1300–1350) has an ode on " Rhaiadr Gwy," and Lewys-Glyn-Cothi, one of the most travelled Welshmen of the fifteenth century, wrote another on the Vale of Ithon. Dr Dee, the Beguildy mathematician, shone further afield, for he was one of the ornaments of Elizabeth's court.

Coming down to the end of the eighteenth century we find Radnor well represented in the literary world by Edward Davies, the author of *Celtic Researches*, and a little later by George Stovin Venables, one of the earliest contributors to the *Saturday Review*. About the same time lived Carnhuanawc, the Vicar of Cwmdu, who, though a Breconshire man, studied the wide field of Radnor antiquities to good purpose and wrote a great deal on the subject.

The best known name of the county in pictorial art is that of Thomas Jones Pencerrig, who painted *The*

*Merry Villagers* and *Dido and Aeneas.* In music it has no outstanding name, but "St Garmon"—one of the most popular hymn tunes of to-day—was composed by J. Price of St Harmon's, and by him was called after his native place.

The late Mr James Mansergh, the engineer of Cwm Elan, was sheriff for the county in 1901, and another great engineer, James Watt, bought Doldowlod to reside in after his great services to British trade and commerce had enabled him to reap his reward.

Of other men of action Radnor can boast of Roger Vaughan, of Agincourt fame, and Bryan Harley of Stanage, who fought so well at Poictiers.

Its most celebrated women also won their renown by their daring, for Ellen the Terrible, whose effigy and story are to be seen in Kington Church, was a member of the Llynwent family of Vaughans. Maud St Valery, better known as "Moll Walbee" of Hay and Llowes, was a kindred spirit, for it was she who had the temerity to charge King John to his face with the murder of Prince Arthur at Rouen. Her husband had been at the Norman town when the dastardly deed was done, and she evidently knew more of the affair than was compatible with John's peace of mind. She was thrown into the dungeon of Windsor Castle, where she and her son were allowed to die of starvation.

Of politicians the greatest Radnor names are Sir David Williams, the friend and associate of the great Burghley of Elizabeth's court ; Sir Rowland Gwynne—Macaulay's honest country gentleman—who introduced the Bill for

Moll Walbee Stone, Llowes Churchyard

settling the succession in the House of Hanover; and Sir George Cornewall Lewes—Lord Palmerston's lieutenant in the fifties and sixties of the last century. There is a bronze statue of the last-named in front of the Hereford Shire Hall, and his native place—New Radnor—has also shown its appreciation of his worth by erecting a column to his memory in its main street.

## 26. THE CHIEF TOWNS AND VILLAGES OF RADNORSHIRE.

(The figures in brackets after each name give the population in
  1901, except in the case of the Urban areas which are those
  of 1911.)

**Abbey-Cwm-Hir** (396), a village six miles north-east of
Rhayader, is called after the celebrated abbey of the same name.
Llewelyn the Last is supposed by many to have been buried here.
(pp. 29, 35, 54, 57, 100, 109, 113, 114, 115.)

**Aberedw** (185), at the junction of the Edw and Wye four
miles south-east of Builth, has many historical associations with
Llewelyn ap Gruffydd. The cave where he is supposed to have
hidden is near. The ruins of the castle are nearly obliterated.
The Aberedw Rocks above the Wye bed are interesting to
geologists. (pp. 23, 31, 70, 96, 110, 118, 121, 131.)

**Beguildy** (799) is a large parish in the north-east of the
county seven miles north-west of Knighton. It has an ancient
camp called Crug-y-Byddar, said to have been occupied by
Caradoc. It is the birthplace of Dr John Dee, the mathematical
instructor of Queen Elizabeth. (pp. 99, 103, 139.)

**Boughrood** (200) is a village in the Wye Valley ten miles
south-east of Builth. It has the remains of an ancient castle that
commanded one of the principal fords of the Wye. (pp. 25, 42,
93, 119, 125, 131.)

**Cascob** (55) is a hamlet on the north-east slope of the Radnor Forest some five miles north-west of Presteign. It is mentioned in Domesday as Cascope. A portion of the parish called Lytton used to form an insulated portion of Hereford. (p. 11.)

**Cefnllys** (137), an ancient parish out of a part of which the modern Llandrindod Wells Urban District has been formed. Prior to 1886 it formed one of the contributory boroughs attached to New Radnor for separate parliamentary representation. (pp. 29, 99, 112, 118, 136, 137.)

**Clyro** (686), on the Wye one mile north-west of Hay, is a very beautiful district rich in remains of antiquity and in manor-houses. (pp. 26, 31, 52, 85, 92, 97, 109, 115, 124, 127, 129.)

**Discoed** (84), a parish in north-east Radnor two and a half miles west of Presteign, intersected by Offa's Dyke. It is mentioned in Domesday as Discote. Near it is Beggar's Bush, an inn said to have been so named by Charles I. (pp. 103, 128.)

**Disserth** (453), a village in the Ithon Valley five miles north-west of Builth. Its "feasts," celebrated in honour of its patron saint, were much attended in the eighteenth century. The Roman road joining Caerfagu to Llechryd passed through this place. (pp. 70, 86, 110, 128, 137.)

**Evenjobb** (254), four and a half miles north-west of Kington, is mentioned in Domesday. Offa's Dyke passes through it. Newcastle Camp, a prehistoric fortress, is in its neighbourhood. (pp. 103, 124.)

**Gladestry** (258) is a pretty village on the river Avon in the east of the county four miles south-east of New Radnor. Gladestry Court is an ancient house. (pp. 16, 54, 93, 110.)

**Glasbury** (460) is a beautiful village situated on the Wye at the extreme south of the county. The Welsh call it *Y Clâs*

(the cloisters).   Maes-yr-onen, Llwynllwyd, and Maesllwch are
noted in the immediate vicinity.   (pp. 9, 10, 11, 13, 26, 35, 49,
52, 54, 57, 83, 93, 110, 124, 138.)

**Glascwm** (314), a village eight miles east of Builth in the
centre of the county, which possesses a famous church mentioned
by Giraldus in 1188.   Clasgwyr, a prehistoric double intrench-
ment, lies near the village.   (pp. 16, 54, 98, 107, 110, 127, 129.)

Dinbod Castle Moat, Llanano

**Knighton** (1886) is a market town on the river Teme in
the north-east corner of the county.   It is very old, its streets are
steep and narrow, and many of its houses quaint.   It used to
have factories of woollen cloth but now they are much decayed,
and the chief trade at present consists in the buying and selling
of live stock.   It has many fairs and a large sheep market.

It was once a town of much importance.   The Welsh know
it as Tref-y-Clawdd, "the town on the Dyke."   Traces of Offa's

D. R.                                                           10

Dyke are still seen in the vicinity and the famous "Caractacus Camp" is on the Shropshire side of the river opposite the town.

Very little is known of its Norman castle. Before 1886 Knighton was contributory to the borough of Radnor. (pp. 9, 10, 20, 49, 52, 78, 82, 87, 88, 93, 103, 121, 124, 131, 135, 136.)

**Knucklas** (181) is a decayed township near Knighton, which has the remains of an ancient castle. Vavasor Powell was a native. (pp. 20, 121, 136, 138.)

**Llanano** (244), on the Ithon ten miles north-west of Rhayader, has the ruins of Dinbod Castle. Ffynnon Newydd at the foot of Rallt is a chalybeate spring. (pp. 110, 112, 114.)

**Llanbadarn Fynydd** (452) is an outlying parish in the extreme north of the county. Some lead-mining was done here in former days. Ffynnon Dafydd-y-Gôf is a sulphur-spring in its neighbourhood. (pp. 29, 109.)

**Llanbister** (704), a large parish in the north of the county. Llynwent, an ancient mansion of the Vaughans, lies within it. (pp. 107, 109, 110, 114, 124.)

**Llanddewi-ystrad-Enny** (391) is the centre of a district noted for its remains of antiquity, of which the Gaer (a camp) Bedd-y-gre (a tumulus), and Cwmaron Castle are the best known. (pp. 35, 96, 100.)

**Llandegley** (292) is an extensive parish five miles north-west of Radnor. It has a sulphur spring at the foot of Rhosgoch Mt. The Llandegley Rocks are of great interest to geologists. (pp. 30, 51, 85, 112.)

**Llandrindod Wells** (2779) is by far the most important and well known of Radnorshire towns, from its mineral springs, which are visited by about 100,000 people during the season. Although these wells, with their sulphur, chalybeate, saline, and magnesium waters, have been known for the last 200 years, it is only within the last quarter of a century that they have attracted the general attention of the country. Until Llandrindod arose there was no convenient centre where North and South Wales could meet for the consideration of their common national interests. Conferences, conventions, and festivals of the combined provinces are now almost invariably held here. Llandrindod was a competitor for the holding of the National Eisteddfod in 1909. (pp. 7, 8, 51, 55, 71, 82, 84, 85, 96, 97, 98, 100, 131, 135, 139.)

**Cockpit on Llanelwedd Rocks**

**Llanelwedd** (432), on the Wye, is a Radnorshire suburb of Builth, and is the busiest quarter of that thriving town. It has railway depôts and some large stone quarries. Near it are the Carneddau Hills, celebrated for their prehistoric remains.

Llanelwedd was the scene of the Builth Historical Pageant in 1909. Like Builth, this town has suffered much from the ravages of fire. It was burned to the ground in 1777. (pp. 13, 24, 35, 41, 55, 58, 77, 80, 84, 85, 94, 97, 98, 100, 101, 107, 123, 128, 131.)

**Llanfihangel-nant-Melan** (126) is a village in the midst of the antiquities and physical beauties of the Radnor Forest. Tomen Castle, a large tumulus, and Water-break-its-neck, a remarkable cascade, are in the immediate neighbourhood. (pp. 33, 128.)

**Llangunllo** (478) is an upland parish in the north-east of the county six and a half miles south-west of Knighton. It

Newbridge-on-Wye School

has several prehistoric remains, chief of which is "The Camp" a circular tumulus. *Cerwyni Cynllo*, "Cynllo's barrels," is a series of remarkable cavities in the river bed. (pp. 19, 103.)

**Llansantffraed-in-Elvel** (230), four and a half miles north-east of Builth in the Forest of Colwyn, was once a Royal Chase. It has a castle, near which Llewelyn-ap-Gruffydd fought an indecisive battle with Edmund Mortimer. (p. 129.)

**Llansantffraed Cwmdauddwr** (1279) is a parish west of Rhayader which contains most of the waters of Cwm Elan. (pp. 54, 84, 86, 93, 97, 98, 115, 120.)

**Llowes** (218) is a suburb of Hay, chiefly noted for its Moll Walbee stone.   (pp. 26, 98, 101, 140, 141.)

**Michaelchurch** (108), in the fertile Vale of Arrow, is a village set in the midst of Roman and prehistoric remains.   The

Interior of Old Radnor Church

camp of "The Gaer" is a large one and was connected by Roman roads with Gobannium (Abergavenny) and Caerfagu.   (pp. 16, 93, 128, 129.)

**Nantmel** (1011), a township between Rhayader and Llandrindod, has largely benefited from the Cwm Elam reservoirs. It has an aqueduct carrying the Birmingham water over the valley.   (pp. 34, 69, 87, 100, 123.)

**Newbridge-on-Wye** (704), in the parish of Llanyre, is a flourishing village in the midst of a large agricultural district. It has several fairs where much business is transacted. Its bridges have often been washed away. A new bridge was opened in 1911. (pp. 22, 23, 30, 78, 79, 98, 129.)

**New Radnor** (405), once the county town, is a township situated on the Somergil in the east of the county. In spite of its name it is a very old town, being mentioned in Domesday and known to have been founded by Harold Godwin. The Crusades were preached in its streets in 1188. It received a charter from Queen Bess and until 1886, with its contributory boroughs, had separate representation in Parliament. It was once a considerable town but at present it has sadly decayed. It still has a Guild Hall, a part of which is utilised as a market. Its castle underwent two sieges, one during the Glyndwr rising and the other in the Civil War. Sir George Cornewall Lewes was a native of the district. A column to his memory was erected in the main street in 1864. (pp. 4, 19, 88, 91, 93, 98, 104, 117, 118, 131, 135, 136, 142.)

**Norton** (283) is a charming village situated in the valley of the Lug. Its neighbourhood is one of the most fertile in the county. On the hillside between this village and Knighton is placed a monument to a former Member of Parliament, Sir Richard Green-Price. (pp. 49, 52, 88, 93, 103.)

**Old Radnor** (343) is an ancient township which has greatly decayed during the last few centuries. Its church is a famous one, and several of the county seats in the neighbourhood are historic. Burlinjobb (pronounced Birchope), mentioned in Domesday as Birchelincope, is a suburb where limestone is quarried, and Burfa is an ancient military camp in the vicinity. (pp. 4, 18, 35, 51, 57, 83, 85, 103, 110, 112.)

**Painscastle** (171), in the southern portion of the shire, is another decayed town. It once had a market and some trade, but at present it is a mere hamlet.    Its castle has a very romantic history.    (pp. 13, 31, 33, 52, 54, 75, 86, 118, 135, 136, 137.)

**Penybont** (292) lies mainly in the parish of Llandegley five miles north-east of Llandrindod.    The Radnorshire Agricultural Society held its meetings here for years, and it was here too that the Radnorshire Bank was first instituted by Mr John Price.

**Pilleth** (109) is a district to the south of Knighton, for ever famous on account of the defeat inflicted on Edmund Mortimer by Rhys Gethin in 1402.    The battlefield is also known as Brynglâs.    A clump of trees was planted some years ago to mark the historic spot.    Shakespeare refers to the battle in *King Henry IV*, and Adam, the chronicler of Usk, said that he heard of it even in Rome.    Pilleth Court and Monachty are old-world mansions of the district.    St Mary's Well was supposed to have miraculous powers.    (pp. 19, 91, 109, 115, 124.)

**Presteign** (1141), on the Lug, is the county town of Radnor. It has had a long history since Martin, Bishop of St Davids in 1293, made it the leading town of Melenydd by giving it a weekly market.    Another bishop, Lee, President of the Court of the Marches, gave it an evil name when he spoke of being "at Presteign, even among the thickest of the thieves."    It has traditions of manufactures in Tudor times, while the parish registers show that it suffered much from the Plague in 1593, 1610, and 1636.    The Guild Hall and the Market Hall are both in Broad Street.    Dr Lucas the divine and John Bradshaw the regicide were both natives.    The Radnor Arms, the most famous hostel of the town, was owned by the latter.    During the eighteenth century the Oak Inn, Broad Street, was notorious for cock-fighting.    Presteign Grammar School, founded by John Beddows, clothier, was a well-known one.    A new Secondary school has recently been opened here by the Radnorshire County Council.    (pp. 6, 9, 20, 49, 52, 58, 75, 82, 85, 87, 93, 103, 109, 124, 127, 128, 131, 135, 136, 139.)

**Rhayader** (1215), on the Upper Wye, is a typical Welsh market town, the centre of a highland agricultural district, and has a large number of fairs. It is also a favourite angling centre.

Rhayader has always borne the name of a lawless town from the erection of its castle by the Lord Rhys in 1178 down to the Rebecca riots of the forties. It is asserted that the Radnor Assizes were formerly held here but were removed to Presteign after the murder of one of H.M.'s judges. Pen-y-maes to the north of the town is pointed out as the spot where convicts were executed of old.

Rhayader has several small tanyards and formerly manufactured cloth. (pp. 8, 11, 13, 21, 22, 28, 29, 34, 39, 43, 73, 78, 82, 83, 84, 85, 86, 93, 94, 95, 96, 114, 120, 121, 123, 126, 127, 128, 129, 130, 131, 134, 135, 136, 137, 139.)

**St Harmon's** (653), a parish to the north of Rhayader, is supposed to have been named after the hero of the Hallelujah Victory, so often referred to in ancient Welsh chronicles. Its church, mentioned by Giraldus, contained the miraculous staff of St Curig which cured glandular diseases of all kinds. This staff is said to have been burnt in the turbulent days of the Reformation. (pp. 20, 96, 98, 107, 140.)

**Stanage** (153), a township near Knighton, is noted now only for Stanage Castle, the home of the Rogers family. On Reeves Hill near the village is a prehistoric camp.

Domesday mentions Stanage as belonging to Osbern, one of the Norman adventurers. (pp. 91, 93, 96, 124, 140.)

**Walton** (162), a hamlet near Old Radnor, is the meeting place of many of the roads of the eastern part of the county. On that account it is a favourite centre for county gatherings. (p. 88.)

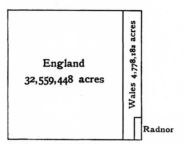

Fig. 1. The Area of Radnorshire (301,165 acres) compared with that of England and Wales

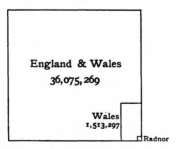

Fig. 2. The Population of Radnorshire (22,589) compared with that of England and Wales, 1911

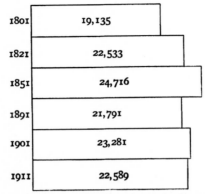

Fig. 3. Table showing Variation in Population
of Radnorshire

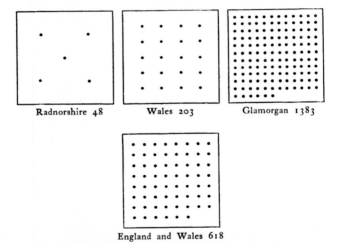

Fig. 4. Comparative Density of Population per Square Mile
in Radnorshire, Wales, Glamorganshire, and England
and Wales in 1911

(*Each dot represents ten persons*)

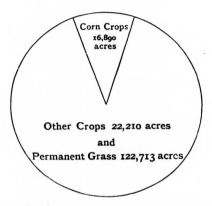

Fig. 5.  Area under Corn Crops in Radnorshire in 1910

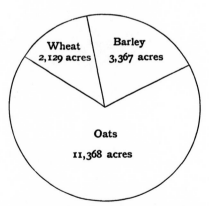

Fig. 6.  Proportionate Areas of Chief Cereals in
Radnorshire in 1910

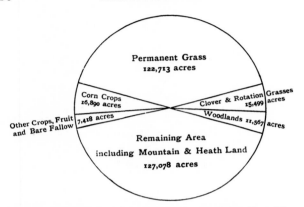

Fig. 7. Proportionate Area of Cultivated and Uncultivated
Land in Radnorshire in 1910

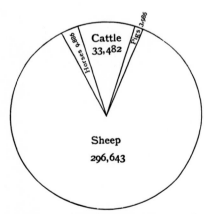

Fig. 8. Proportionate numbers of Live Stock in
Radnorshire in 1910

Lightning Source UK Ltd.
Milton Keynes UK
UKOW052251080213

206048UK00001B/12/P